T0189737

Modern Accelerator Technologies
for Geographic Information Science

Xuan Shi • Volodymyr Kindratenko
Chaowei Yang

Editors

Modern Accelerator Technologies for Geographic Information Science

 Springer

Editors
Xuan Shi
Department of Geosciences
University of Arkansas
Fayetteville, AR, USA

Volodymyr Kindratenko
Department of Electrical
  and Computer Engineering
University of Illinois
Urbana, IL, USA

Chaowei Yang
Department of Geography
  and GeoInformation Sciences
George Mason University
Fairfax, VA, USA

ISBN 978-1-4899-7861-5   ISBN 978-1-4614-8745-6 (eBook)
DOI 10.1007/978-1-4614-8745-6
Springer New York Heidelberg Dordrecht London

Contents

Part I
Introduction

Chapter 1
Modern Accelerator Technologies for Geographic Information Science

Xuan Shi, Volodymyr Kindratenko, and Chaowei Yang

Keywords Modern Accelerator Technologies • GIScience

Geographic Information System (GIS) enables heterogeneous geospatial data integration, processing, analysis, and visualization. With a variety of software tools, GIS makes substantial contribution to the advancements of science, engineering and decision-making in geospatial-related natural and social sciences, public safety and emergency response, spatial intelligence analytics and military operations, ecological and environmental science and engineering, and public health. Geospatial data represents real-world geographic features or objects using either vector or raster data models. In the vector model, features are captured as discrete geometric objects and represented as points, lines or polygons with non-spatial attributes. In the raster model, features are represented on a grid, or as a multidimensional matrix, including satellite imagery and other remotely sensed data.

When geospatial data is increasingly available along with an accelerating increase in data volume, it has been a grand challenge to manipulate large-scale data and complete data processing and analytics using traditional GIS software tools. Emerging computer architectures and advanced computing technologies provide a promising solution to employ massive parallelism to achieve scalability with high performance for data intensive computing over big geospatial data.

X. Shi (✉)
Department of Geosciences, University of Arkansas, Fayetteville, AR 72701, USA
e-mail: xuanshi@uark.edu

V. Kindratenko
National Center for Supercomputing Applications, University of Illinois
at Urbana-Champaign, Urbana, IL 61801, USA

C. Yang
Department of Geography and GeoInformation Science, College of Science,
George Mason University, Fairfax, VA 22030, USA

X. Shi et al. (eds.), *Modern Accelerator Technologies for Geographic Information Science*,
DOI 10.1007/978-1-4614-8745-6_1, © Springer Science+Business Media New York 2013

The GIScience community has been an active user of high-performance computing (HPC) resources for both data-intensive and compute-intensive applications. The HPC technology, however, is shifting from homogeneous systems employing identical processing elements to hybrid computing architectures that employ multi-core and many core processors in combination with special-purpose chips. New multi-core architectures combined with application accelerators hold the promise of increasing performance by exploiting levels of parallelism not supported by the conventional systems. This book is a response to the research needs and has accommodated 17 papers in 5 parts, each covers a specific category.

Three chapters in Part II provide *an overview of Modern Accelerator Technologies (MAT) for scientific computation and geoscience applications*, including Graphics Processing Units (GPUs), Intel's Xeon Phi coprocessor, and cloud computing infrastructure. Today AMD and NVIDIA GPUs and Intel's Xeon Phi coprocessor are the leading computational accelerators. While GPU-based accelerators have been around since 2007, Xeon Phi is a newcomer with up to 61 Intel Architecture general purpose processor cores plus a powerful, new wide vector processing unit in each core. Jeffers' paper provides an overview on the Xeon Phi architecture and how it can support developing highly parallel, data intensive applications in geosciences. Choi and Vuduc's work gives a brief history and overview of modern GPU systems, including the dominant programming model that has made the compute capabilities of GPUs accessible to the programming-literate scientific community. This chapter offers examples to discuss GPU hardware and programming environment and the main principles as well as the new features in NVIDIA's GPUDirect technology. Cloud computing is considered as the next generation computing platform with the potential to address the computing challenges and redefine the possibilities of geoscience and digital Earth. Huang et al. introduces through examples how cloud computing can help accelerate geocomputation by elastically integrate and deliver most advanced computing technologies including HPC, GPU, and others.

Part III presents several chapters on the use of *MAT in GIScience applications*. Zhang reports on designing and implementing a spatial join algorithm, fundamental to spatial databases and vector GIS, on GPUs by using generic parallel primitives. The results demonstrates that when natively implemented on GPUs, such operations can be speed up significantly close to ten times. Li et al. describes an effort to design and implement a GPU-based visualization pipeline for 5D geospatial data stored in scientific data formats. The developed visualization pipeline runs almost entirely on a GPU. The authors also discuss advantages and disadvantages of employing GPU and CPU alternatively for geovisualization applications. Medrano and Church present a new breadth-first-search parallelization of the Near-Shortest Path algorithm used to generate route alternatives. The authors introduce a parallel efficiency measure of how successful their parallelization is when distributing the workload among multiple computational nodes. Franklin et al. present High-dimensional Overdetermined Laplacian Partial Differential Equations algorithm and its implementation for lossy compression of high-dimensional arrays of data. The authors developed a MATLAB-based implementation of the proposed algorithm that uses GPUs to speed-up the execution of computationally intensive compression tasks. Tang identified and discussed the fundamental aspects when using GPUs to accelerate agent based models (ABMs) including random number generation, parallelization of agent-based interactions, analysis of agent and environment patterns, and evaluation of computing performance. He used a case study of modeling spatial opinion exchange to illustrate the massively parallel computing power of GPUs for accelerating agent-based modeling.

In Part IV several authors report on the success with *MAT in remotely sensed data processing and analysis* domain. Bin et al. describes an effort to achieve a high-speed multi-velocity-channel processing of Costas signal pulse compression for high speed radar signal processing. The authors use a parallel processing scheme and propose a GPU-based implementation that delivers over two orders of magnitude performance improvement over traditional computing platforms allowing deal with large-scale radar signal processing tasks. The last two articles deal with image processing algorithms used for image segmentation and classification. Ye and Shi report on the results of parallelizing ISODATA algorithm for unsupervised image classification used in remote sensing applications. The authors describe in great details parallelization and optimization strategies employed on the GPU and compare classification results to a widely-used remote sensing software. Huang et al. also employ GPUs for speeding up mean shift segmentation algorithm used in remote sensing applications. The authors also consider a hybrid architecture in which both CPU and GPU are involved in the computation. As indicated by the articles presented in this Chapter, image and signal processing problems are particularly suitable for massively-parallel architectures, such as GPUs, and their performance can be greatly enhanced by re-implementing key computational components on such architectures.

Part V introduces *how multi-core technology can accelerate geospatial service computation and spatial statistical calculation*. Web Map Tile Service (WMTS) has been increasingly adopted in many online mapping services and applications. In practice, WMTS scalability has been a concern when WMTS servers handle massive concurrent requests. When client users increase dramatically, the torrent of client requests places overwhelming pressure on the web server where the WMTS is deployed, causing significant response delay and serious performance degradation of the WMTS. In this chapter, Wu et al. introduced how to build a high performance cluster (HPC) to handle large scale service request. In another paper, Li et al. introduced how cluster-based caching systems can accelerate users' access to large-scale network services. How to configure numerous parameters to make cluster-based caching servers cooperate with each other to efficiently share cached data is critical to obtain optimal performance. This paper analyzes tile access characteristics in Web GIS applications and simulates cluster-based caching system through a trace-driven experiment based on the log files. The configuration of each parameter in a cluster-based caching system is quantitatively analyzed to obtain a global optimal combination of parameters. In the last paper in this Chapter, Laura and Rey introduce an improved parallel optimal choropleth map classification algorithm to support spatial analysis. This work contributes to the development of a Distributed Geospatial CyberInfrastructure and offers an implementation of the Fisher-Jenks optimal classification method suitable for multi-core desktop environments.

Part VI includes the visions and future research needs of utilizing accelerating technologies for different domains. Ye and Shi discuss how to use high performance computing to speedup the spatiotemporal interaction and analyses of crime and police activities for future potential prediction of such social events based on historical records and social sciences. Guan and Shi introduce how urban land use simulations could be accelerated using latest accelerator technologies including CPU, GPU, and High performance computing. Following the foundational steps, they discuss the research challenges and opportunities of using the accelerating technologies to address the challenges. Wang introduces the challenges in spatially integrating environmental systems models to address more complex and broader geographic scope of environmental problems. Future research directions for using accelerating technologies for integrated environmental modeling are discussed at the end of the chapter.

At last, we would like to thank all authors for their excellent works and contribution to this initiative. Specifically we want to thank Springer editors for their patience, encouragement and continuous support to make this book successful.

Part II
Overview of Modern Accelerator Technologies (MAT) for Scientific Computation

Chapter 2
A Brief History and Introduction to GPGPU

Richard Vuduc and Jee Choi

Abstract Graphics processing units (GPU) are increasing the speed and volume of computation possible in scientific computing applications. Much of the rapidly growing interest in GPUs today stems from the numerous reports of 10–100-fold speedups over traditional CPU-based systems. In contrast to traditional CPUs, which are designed to extract as much parallelism and performance as possible from sequential programs, GPUs are designed to efficiently execute explicitly parallel programs. In particular, GPUs excel in the case of programs that have inherent *data-level parallelism*, in which one applies the same operation to many data simultaneously. Applications in scientific computing frequently fit this model. This introductory chapter gives a brief history and overview of modern GPU systems, including the dominant programming model that has made the compute capabilities of GPUs accessible to the programming-literate scientific community. Its examples focus on GPU hardware and programming environment available from a particular vendor, NVIDIA, though the main principles apply to other systems.

Keywords CUDA • GPGPU • High-performance computing • Parallel programming • SIMD • Vector processors

2.1 A Brief History

The broader adoption of GPUs today stems from advances in programming and in the use of power-efficient architectural designs. We briefly recount these advances below.

R. Vuduc (✉) • J. Choi
School of Computational Science & Engineering, College of Computing,
Georgia Institute of Technology, Atlanta, GA 30332, USA
e-mail: richie@cc.gatech.edu; jee@gatech.edu

X. Shi et al. (eds.), *Modern Accelerator Technologies for Geographic Information Science*,
DOI 10.1007/978-1-4614-8745-6_2, © Springer Science+Business Media New York 2013

2.1.1 From Special-Purpose to General-Purpose Programming

GPUs were originally designed to accelerate graphics related tasks such as texture mapping or vertex shading commonly used in computer games and 3D graphics to render images. During the early 2000s, to increase the realism of 3D games and graphics, GPU designs were enhanced to deliver more *polygons per second* of performance. This style of geometric computation is heavily floating-point intensive, and the raw throughput of such operations vastly outstripped the capabilities of conventional, albeit general-purpose, CPU processors. Lured by these capabilities, developers of scientific software began looking for ways to exploit GPUs beyond the graphics rendering tasks for which they were designed.

Early GPUs were hard to program for anything other than graphics applications. The hardware consisted of fixed graphics pipelines; to program them, one had to use specific application programming interfaces (API) such as OpenGL and DirectX, or shader languages such as C for Graphics (Cg). (Regarding terminology, we will refer to languages and/or libraries needed to program a machine as *programming models*.) If one wished to program something other than graphics, it was difficult to do so since every operation had to be mapped to an equivalent graphics operation.

There were numerous efforts in the research community to ease GPU programming through high-level language extensions (Michael et al. 2002; Ian Buck et al. 2004). This research was the basis for NVIDIA's Compute Unified Device Architecture (CUDA), a more general-purpose computing platform and programming model, which is arguably the dominant albeit largely vendor-specific model available today. CUDA allows developers to use C (as well as many other languages, APIs and directives) as a high-level programming language for programming NVIDIA GPUs. Uptake of CUDA within the scientific research community has been reasonably swift, with approximately 37,000 published research papers on the use of CUDA and 1.6 million CUDA downloads as of 03/19/13. Arguably, CUDA and its ilk have turned an otherwise special-purpose platform into one suited to general-purpose use, hence the term *general-purpose GPU* (GPGPU) computing.

2.1.2 Single-Instruction Multiple-Data Designs

Concomitant with the advances in programming models were advances in the hardware design itself. A key idea in GPU architectures is *single instruction multiple data* (SIMD) design, a type of parallel computer architecture in which a single instruction prescribes that the machine perform a computational operation on many words of data simultaneously (Flynn 1972). The first SIMD computer was the ILLIAC-IV, built in the late 1960s (Bouknight et al. 1972). It was later followed by other systems including ICL's Distributed Array Processor (DAP) (Reddaway 1973). However, it was not until the 1980s that interest in SIMD computers peaked and led to the development of Connection Machine CM-1 (Hillis 1982) and MasPar

MP-1 (Blank 1990). Many of these systems had common architectural features: there was typically one or more central processing units or sequencers that fetched and decoded instructions; instructions were then broadcast to an array of simple, interconnected processing elements.

Of these systems, Connection Machine CM-1 perhaps most closely resembles modern GPUs. The CM-1 system consists of four processor arrays, each consisting of 16,000 processors, and from one to four front-end computers depending on how the processor arrays are used; the arrays can be used separately, in pairs, or as a single unit. The front-end computers control the arrays and issue instructions to them. The CM-1 also provides a virtual number of processors to fit application needs; then, the physical processors are time-sliced over multiple data regions that have been assigned to it, enabling effective and flexible programming and utilization of the available processors. As we will see later, this virtual processor abstraction closely resembles modern GPGPU programming.

Modern GPUs have a come a long way since the days of CM-1 and MasPar. For under $500, one can now purchase a desktop GPU that can execute 3.5 trillion floating-point operations per second (teraflop per second, or TFLOP/s) within a power footprint of just 200 Watts. The basic enabling idea is a SIMD execution style.

2.2 Overview of GPGPU Hardware and Programming Today

The CUDA programming model, introduced by NVIDIA in November 2006, simplifies how one may express data parallel programs in a general-purpose programming environment. CUDA originally extended the C++ language, with recent additional support for Fortran and Java. This chapter summarizes the key C++ extensions; we encourage interested readers to the latest CUDA Programming Guide for more details (NVIDIA 2013).

2.2.1 A GPU-Based System

Figure 2.1 shows a typical GPU system. It consists namely of one or more GPU *devices*, which is connected to a *host* system. The host system is a conventional computer consisting of one or more general-purpose CPUs and a communication channel between them and the GPU(s). On desktop and workstation systems, this channel is typically a PCI Express (PCIe) bus.

This host-device design implies an *offload* model of execution. In particular, an application begins running on the host and then offloads computation to the GPU device. Moreover, the programmer must copy data from the host to the GPU device and copy any needed results from the device back to the host as needed. The host and device may otherwise execute simultaneously (and asynchronously), as permitted by the application.

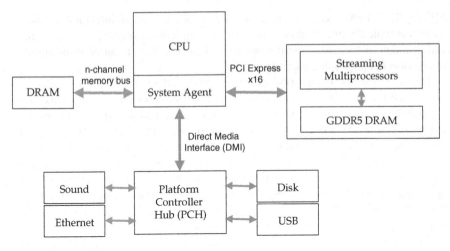

Fig. 2.1 A typical GPU system

This system design has both performance and programming implications. Regarding performance, an application will only see an improvement from use of the GPU if the data transfer time is small relative to the speedup from using the GPU in the first place. Regarding programming, the programmer will write both host code and device code, that is, code that only runs on the host and separate code that may only run on the device; he or she must also coordinate data transfer as the application requires.

Briefly, the typical steps involved in writing a CUDA program may be as follows:

1. Allocate and initialize input and output data structures on host.
2. Allocate input and output data structures on device.
3. Transfer input data from host to device.
4. Execute device code.
5. Transfer output data from device to host.

CUDA provides an API for manipulating the device (e.g., allocating memory on the device) and a compiler (*nvcc*) for compiling the device code.

2.2.2 GPU Architecture

Before we can go into the details of the CUDA programming model, it is necessary to talk about the GPU architecture in more detail. This is because the CUDA programming model (as well as the other models) closely reflects the architectural design. This close matching of programming model to architecture is what enables programs that can effectively use the GPU's capabilities. One implication is that although the programming model abstracts away some details of the hardware, for a program to really execute efficiently, a programmer must understand and exploit a number of hardware details.

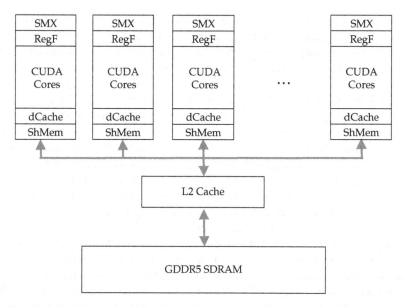

Fig. 2.2 Diagram of a basic GPU architecture

Figure 2.2 shows a basic GPU architecture. First, the GPU has both processing units, which perform actual computation, and a main memory, which store all of the data on which the GPU may operate. The processing units are labeled *Streaming Multiprocessors* (SMX) in the figure, and the memory is labeled *SDRAM*. (The *GDDR5* designation refers to the particular type of memory and protocols used when transferring data).

The SMX units are further divided into several components. At the heart of an SMX is a number of small *functional units* called "CUDA cores." One may regard these CUDA cores as the basic hardware for performing primitive operations, such as floating-point operations.

These CUDA cores share a number of other hardware resources, the most important of which is the *register file* (RegF). A CUDA core may only perform primitive operations on data stored in registers. Roughly speaking, variables (i.e., "local variables") that appear in a CUDA program are stored in registers. The compiler, when it translates the CUDA program into machine code, manages the use of these registers and the mapping of program variables to registers. As such, they are not really exposed to the programmer. However, to achieve good performance, a programmer should keep in mind that since registers are directly connected to the cores, any data stored in registers will be much faster to access than data stored elsewhere in the system.

Indeed, this basic principle underlying register use applies more generally to the entire *memory hierarchy* of the GPU. The memory hierarchy refers to the collection of memories including the main memory (SDRAM), the registers, and a number of intermediate memories. The relative speed of accessing data in registers may be 100× or 1000× faster than doing so from main memory. However, the capacity of all the

GPU's registers compared to main memory may also differ by that same factor. For instance, the aggregate register capacity across all SMX units typically numbers in the millions of bytes, or megabytes, MB; by contrast, the capacity of the main memory is much larger, numbering typically in the billions of bytes, or gigabytes (GB). Therefore, using the memory hierarchy effectively is critical to achieving speedups in practice.

The hierarchy refers to additional intermediate staging levels of progressively smaller but faster memories between main memory and the register file. In Fig. 2.2, these are the *caches* (i.e., the so-called level-1 or *dCache* and the so-called level-2 or *L2 Cache*) and the *shared memory (ShMem)*. Caches are managed automatically by the hardware: the caching hardware "observes" the data requests from the processing units and try to keep frequently accessed data in the cache. By contrast, the shared memory is programmer-managed: the programmer controls exactly what data is in the shared memory at any point in time during the computation.

There are some additional useful details to know about the memory hierarchy. The CUDA cores on a given SMX share the dCache; all SMX units on the GPU share the L2 Cache. Regarding their relative capacities, the CUDA cores of a given SMX share the register file (typical capacity is 256 KB). The total ShMem and dCache capacity is 64 KB per SMX, where the ShMem and dCache may actually be re-configured by the programmer to be either 16 KB of programmer-managed memory with 48 KB of dCache, or 48 KB of programmer-managed memory and 16 KB of dCache, or split equally between the two on current systems.

More concretely, Table 2.1 shows the exact specifications for the latest NVIDIA GPU, the GTX Titan (released in late 2012). The sheer number of CUDA cores is what makes GPUs so compute-rich. Exploiting this relatively large amount of parallelism—thousands, compared to tens on conventional processors—requires a "non-traditional" programming model.

2.2.3   SIMT and Hardware Multithreading

To fully utilize the available CUDA cores on a GPU, CUDA adopts a variation of SIMD, which NVIDIA refers to as the *single instruction multiple thread* (SIMT) style.

In SIMT, a program consists of a number of threads and all threads execute the same sequence of instructions. Therefore, if all threads execute the same instruction at the same time, just on different data per thread, a CUDA program would simply be a sequence of SIMD instructions.

However, SIMT generalizes SIMD in that it allows individual threads to execute different instructions. This situation occurs when, for instance, threads simultaneously execute a conditional (e.g., "if" statement) but execute different branches. When this occurs, threads are said to *diverge*. Although this behavior is allowed by the model, it has an important performance implication. When threads diverge, their execution is serialized—that is, the threads that take one branch may execute first while the other threads idle until the first threads complete. In other words, SIMT allows flexibility in programming at the cost of performance. For example, the only way to achieve the peak performance as calculated in Table 2.1 is to have absolutely no divergent threads in the code.

Table 2.1 Specifications for the NVIDIA GTX Titan

Parameters	Values
SMX	14
32-bit (Single precision) CUDA Cores	2,688 (192 cores/SMX)
64-bit (Double precision) CUDA Cores	896 (64 cores/SMX)
Clock	837 MHz
Boost clock	876 MHz
Memory interface width	384 bits
Memory clock	6.008 GHz [a]
Single precision peak performance	4709.38 GFLOP/s [b]
Double precision peak performance	1569.79 GFLOP/s
Peak bandwidth	288.38 GB/s [c]

[a] Effective clock rate since DDR memory reads from both rising and falling edges of the signal
[b] Peak Performance = (number of cores) × (boost clock) × 2 FLOP/cycle (fused multiply-add)
[c] Peak Bandwidth = (memory interface width) × (memory clock)

In order to have many CUDA cores, a key design decision in GPUs is to also use simpler cores, compared to their equivalents in CPU systems. Simpler cores tend to be smaller and more energy-efficient, but also slower *per instruction*. However, the long latencies associated with these slower instructions—as well as the relatively slower memory operations—may be mitigated by allowing many simultaneous instructions to be in-flight at any time, with the processors juggling among available instructions. That is, at any given time, a GPU keeps the context of a large number of threads on the chip so that it can switch from one set of threads to another quickly whenever threads are ready to execute. This approach is referred to as *hardware multithreading.*[1]

Indeed, hardware multithreading is feasible because of the SIMT approach, which enables (or rather, *requires*) the programmer to express a large amount of thread-level parallelism. On the hardware side, generous amounts of on-chip resources, such as the large register file and shared memory, are also necessary. Although conventional CPUs also employ hardware multithreading (e.g. Intel's hyperthreading technology), the number of in-flight threads is much smaller—compare 2 in-flight threads per core using Intel's hyperthreading technology, compared to, say, 64 threads per SMX on a representative GPU.

2.3 CUDA

Here we go into the specifics of the CUDA programming model. We will first cover three key concepts: the *thread hierarchy*, the *memory hierarchy*, and *synchronization*. Then, we will discuss commonly used strategies for optimizing CUDA programs for performance. All hardware specifications mentioned in this section will be that of the GTX Titan which is based on the latest generation of NVIDIA's GPU architecture, the *GK110*.

---

[1] It also explains why advocates of GPU design refer to GPUs as being especially suited to *throughput-oriented* execution, rather than *latency-oriented execution* as in CPUs.

Grid

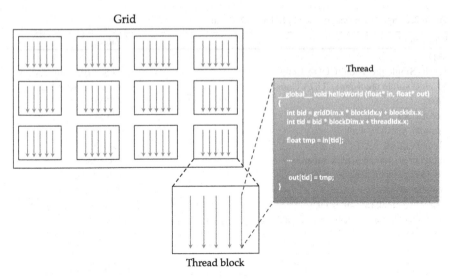

Fig. 2.3 CUDA thread hierarchy

2.3.1 Thread Hierarchy

There are three layers of hierarchy for CUDA threads; threads, thread blocks, and grids. The relationship between these three layers are shown in Fig. 2.3.

Recall that the smallest granularity of computation on a GPU is a *thread*, which simply executes a sequence of instructions, or *kernel*. A programmer may think of the kernel as the "program" that all threads execute simultaneously. While executing the kernel code, each thread has a unique integer identifier that it can use to determine whether to perform thread-specific operations, such as loading a particular data element. A typical CUDA program consists of anywhere from thousands to millions of these threads.

Threads are grouped into *thread blocks*. A thread block may contain any number of threads up to some limit, which on current hardware is 1,024 threads per block. Importantly, thread blocks execute completely *independently*. That is, once a thread block begins execution, it runs to completion; moreover, there is no guarantee on what order thread blocks will execute, so a programmer should not rely on this fact. Additionally, the thread block may be logically configured as a one-, two-, or three-dimensional array of threads. The purpose of a multi-dimensional configuration of threads is to allow for easier mapping of threads to work.[2]

---

[2] For instance, if a GPU kernel operates on a 2-D image, it is very likely most "natural" to assign one thread to perform a computation on each pixel; in this case, a 2-D logical organization of threads is likely to be a sensible way of mapping threads to work.

The significance of a thread block is scalability. When threads are assigned to a multiprocessor, they are done so in the granularity of thread blocks. Threads within a thread block may coordinate their work, but thread blocks—being executed independently as noted previously—may not. Additionally, a thread block completes only when every thread in the block has finished its work. This behavior permits scaling a workload with sufficiently many thread blocks onto GPUs that may have differing numbers of SMXs.

Thread blocks may be further logically organized into *grids*. That is, a grid is a collection of thread blocks, also configurable as a one-, two-, or three-dimensional array. Grid size and dimensions are dictated typically by either the work-thread block mapping, or the total number of thread blocks. A grid can be seen as the entire collection of CUDA threads that will execute a given kernel.

2.3.2 Memory Hierarchy

As noted in the architectural overview above, GPUs have multi-level memory hierarchy. This hierarchy is similar to traditional CPUs; on current generation GPUs, there is a level-2 (L2) cache that is hardware-managed and shared between all multiprocessor cores on the GPU and a level-1 (L1) cache that is local to each core. On current GPUs, the L2 cache is 1,536 KB in size, and all data accesses through the SDRAM, whether read or write, is stored in the L2 cache. The L1 cache is reserved for local memory usage such as register spills or local arrays that are either too large or cannot be indexed with constants.

Additionally, there is also the programmer-managed shared memory. The hardware used to implement shared memory is identical to that of the L1 cache and together they make up 64 KB in size. As needed, shared memory can be configured to be 16, 32, or 48 KB, with the L1 cache taking up the rest. Shared memory is local to a thread block, and is a limited resource that can constrain the number of thread blocks that can be resident on the multiprocessor at the same time.

There is also a 48 KB read-only data cache called constant memory. This is part of the *texture* unit that is necessary for graphics performance, but is also available for general purpose use. The availability of constant memory relieves pressure on the shared and L1 cache by reducing conflict misses.

Lastly, GPUs have large register files to support hardware multithreading. The size of the register file is 256 KB on the GTX Titan and each thread can use as many as 255 32-bit registers.

One interesting point to note is that unlike traditional CPUs where closer you get to the core, the smaller the cache becomes, GPUs show an opposite trend. That is, when going from L2 to L1 to register, the size of the cache goes from 1,536 KB to 1,568 KB to 3,584 KB.

From a programming perspective, CUDA exposes the shared memory to the programmer. That is, the programmer controls completely how data is placed in shared memory. By contrast, the compiler controls registers (though the programmer may influence their use) and the hardware manages the caches.

2.3.3 Synchronization

We noted previously that threads within a thread block may coordinate their activity. The mechanism for doing so is shared data access and synchronization.

With respect to shared data access, threads within a thread block all "see" the same shared memory data. Therefore, they may cooperatively manage this data. For instance, if the thread block needs to perform computation on some chunk of data, each thread may be assigned by the programmer to load a designated piece of that data.

With respect to synchronization, threads may "meet" at a particular program point within the kernel. This operation is called a *barrier*—when a thread executes a barrier, it waits until all other threads have also reached the barrier before continuing. For example, a common pattern is for threads to cooperatively load a chunk of data from main memory to shared memory as noted above, then issue a barrier. By doing so, the threads ensure that all the data is loaded before any thread proceeds with computation.

Importantly, this type of synchronization is really only possible within a thread block, but not between thread blocks, since there is no guarantee on how thread blocks may be scheduled. The only way to do global synchronization across thread blocks is to do so on the host: the host is what launches the GPU kernel, and the CUDA API makes a host function available to check whether all thread blocks have completed.

2.3.4 Performance Considerations

2.3.4.1 Bandwidth Utilization

GTX Titan boasts 288.4 GB/s of bandwidth, an order of magnitude higher than most CPUs, due to its high memory clock frequency and wide interface width of its GDDR5 SDRAM. However, in order to fully utilize the available bandwidth, certain precautions must be taken on how the data is loaded from the memory.

By a principle known as *Little's Law* (from queuing theory), in order to saturate the memory system, the number of in-flight memory requests must be approximately equal to the memory bandwidth multiplied by the latency between the SDRAM and the multiprocessor core. Latency today is typically in the range of 400–800 clock cycles. Consequently, a GPU kernel in practice needs to have several tens of thousands of bytes of data being requested simultaneously. A computation that cannot have that level of memory-level parallelism will not make the most efficient use of the GPU's memory transfer capability.

Another consideration is to ensure that loads from GPU memory (either to shared memory or to registers) are *coalesced*. Conceptually, the GPU memory system is designed to load chunks of consecutive data at a time. Furthermore, the hardware typically schedules consecutively numbered threads at the same time.

Therefore, a good CUDA programming practice is to ensure that consecutively numbered threads that are performing load or store operations do so to consecutive locations in memory. An easy way to make sure data is coalesced is to *stream* the data where threads accesses one or more words of data consecutively.

2.3.4.2   Core Utilization

To understand how to maximize the use of CUDA cores on a GPU, we must briefly sketch how threads are executed on a GPU.

Although threads begin and complete on the multiprocessor cores of a GPU at the granularity of thread blocks, execution of threads occurs at a smaller granularity. This minimum unit of execution is called a *warp*. (On more classical vector processors, a warp is analogous to the vector or SIMD width.) Although the size of a warp can vary across different GPUs, the warp size has always been 32. This value will most likely increase in the future as the number of cores continues to increase with each successive generation.

On a current generation GTX Titan, there are 4 *warp schedulers* on each multiprocessor. Each warp scheduler can take a warp of threads from the same or different thread blocks currently residing on the multiprocessor and issue one or two instructions, depending on the availability of CUDA cores and the number of independent instructions in each warp, to 32 or 64 CUDA cores respectively. Since there are 192 CUDA cores in each multiprocessor, at least 6 sets of instructions need to be issued *every* cycle in order to fully occupy the CUDA cores. This is equivalent to 2 warp schedulers issuing 1 instruction each, and 2 warp schedulers issuing 2 instructions each. This means that at least 2 of the warps need to have 2 independent instructions that can be scheduled simultaneously. Having only data parallelism is therefore not enough to achieve peak performance on the GTX Titan; instruction level parallelism is also required.

2.3.4.3   Special Functional Units

GPUs also have special functional units (SFU) that implement fast approximate calculation of transcendental operations (e.g., trigonometric functions, exponential and log functions, square root) on single precision data. There are 32 of these units in each multiprocessor of a GTX Titan. They can significantly improve performance when needed.

2.4   Advanced Features of CUDA

In this section, we list some of the newer and more advanced features that have been added to the latest versions of CUDA and GPU hardware, version 5.x and compute capabiilty 3.x respectively at the time of this writing (NVIDIA).

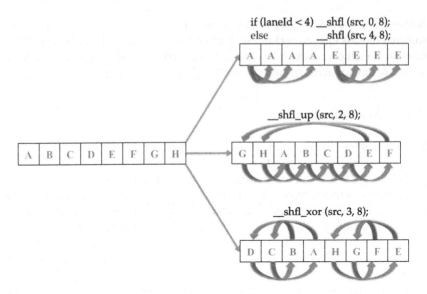

Fig. 2.4 Examples of the shuffle instruction

2.4.1 Shuffle

Shuffle instructions have been added in the latest generation of GPUs to enable data exchange between threads in a warp. The purpose of the shuffle instructions is similar to that of shared memory. However, shuffle instructions have certain performance advantages.

First, sharing data over shared memory requires two steps, a store and a load, whereas a shuffle instruction requires only one, reducing the total number of operations. Secondly, using shuffle instructions will reduce the shared memory footprint for each thread block, which may result in more thread blocks fitting into each multiprocessor and a subsequent increase in performance. On the downside, shuffle instructions are limited to moving 4 byte data types (8 byte data types require 2 shuffle instructions) and cannot be used for inter-warp communication even within the same thread block.

GPUs support three types of shuffle instructions: arbitrarily indexed, shift left/right by n, and XOR. Examples of these three types of shuffle instructions are illustrated in Fig. 2.4.

Readers should keep in mind that although in theory each thread can read from any variable of any other thread in the same warp, doing so using conditionals will create divergence, as exemplified in the topmost shuffle instruction in Fig. 2.4. Therefore, users are recommended to read from the same variable and use computable offsets rather than conditionals whenever possible.

2.4.2 Dynamic Parallelism

Traditionally, all CUDA programs were initiated from the host. After each kernel was launched, the host had to wait for it to complete before launching the next kernel, effectively creating a global synchronization between kernels. This can be expensive as well as being redundant.

In the latest generation of GPUs, CUDA programs can initiate work by themselves without involving the host. This feature, called *dynamic parallelism*, allows programmers to create and optimize recursive and data-dependent execution patterns, opening the door to more varieties of parallel algorithms running on GPUs.

Using dynamic parallelism has other benefits. The ability to create and launch new kernels inside another kernel allows the programmer to change the granularity of the computation on the fly depending on the results of the current kernel execution, possibly increasing the accuracy or the performance of the program as a whole. Finally, it also has the benefit that the CPU resource can now be left uninterrupted for other computation, allowing for better resource utilization and heterogeneous computing.

2.4.3 Hyper Q

One potential problem of having so many cores on a single device is utilization. Unless there is enough work, utilizing the GPU to its full potential can often prove to be difficult or impossible. GPUs provide *streams*, where a single stream is a series of dependent kernel executions, to allow concurrent execution of different kernels to better utilize the available resources. However, in older GPUs it suffered from limitations such as false dependencies due to its *single* hardware work queue which limited the concurrency that could be exploited.

With the introduction of Hyper-Q in the latest generation of GPUs, streams can now run concurrently with minimal or no false dependencies. Figure 2.5 shows an example of concurrent streams with and without false dependencies.

2.4.4 Grid Management Unit

The grid management unit (GMU) is a new feature that allows both CPUs and GPU kernels to launch a grid of thread blocks for execution. It uses *multiple* hardware work queues to paralyze different threads of execution to run concurrently on the same GPU, allowing true concurrency with little or no false dependencies. The use of GMU is what allows both dynamic parallelism and Hyper-Q to work on the latest generation of GPUs.

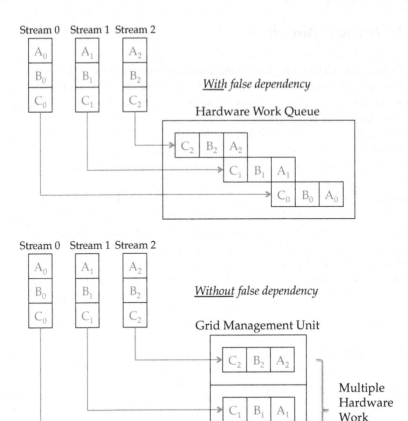

Fig. 2.5 CUDA streams with and without false dependency

2.4.5 GPUDirect

When working with large amounts of data distributed over many GPUs and nodes, inter-GPU or inter-node communication can become a bottleneck for performance. NVIDIA's GPUDirect technology introduces two features that improve performance.

First, GPUDirect allows third-party devices such as Infiniband routers and network interface cards (NIC) to directly access the GPU memory without involving the CPU. This is achieved by having the GPU and the third-party device share the same memory space on the host which removes the need for the CPU to copy the same data from one memory location to another. Secondly, GPUDirect allows point-to-point (PTP) communication between GPUs on the same PCIe bus by communicating directly over the PCIe instead of the host memory. GPUDirect also allows the use of direct memory access (DMA) to communicate, effectively eliminating CPU overheads such as latency and bandwidth bottlenecks.

2.5   Conclusion

In this chapter, we covered the basics of the GPU architecture and its programming model Although this chapter will allow beginners to understand the basic concepts behind CUDA and how it differs from traditional programming, it is far from being a comprehensive guide for GPGPU programming. CUDA and GPU technology continue to evolve even now, requiring constant study and practice in order to effectively utilize the latest GPUs' capabilities and features. Readers are recommended to study the most recent CUDA programming manual and other optimization guides, as well as the latest research papers and various other resources available on the internet in order to have a complete and full understanding of how to use CUDA to leverage the full potential of GPUs.

References

T. Blank. The maspar mp-1 architecture. In *Compcon Spring '90. Intellectual Leverage. Digest of Papers. Thirty-Fifth IEEE Computer Society International Conference.*, pages 20–24, 1990.

W. J. Bouknight, S.A. Denenberg, D.E. McIntyre, J. M. Randall, A.H. Sameh, and D.L. Slotnick. The illiac iv system. *Proceedings of the IEEE*, 60(4):369–388, 1972.

Ian Buck, Tim Foley, Daniel Horn, Jeremy Sugerman, Kayvon Fatahalian, Mike Houston, and Pat Hanrahan. Brook for gpus: stream computing on graphics hardware. *ACM Trans. Graph.*, 23(3):777–786, August 2004.

W. Daniel Hillis. New computer architectures and their relationship to physics or why computer science is no good. *International Journal of Theoretical Physics*, 21(3–4):255–262, 1982.

M. Flynn. Some computer organizations and their effectiveness. *Computers, IEEE Transactions on*, C-21(9):948–960, 1972.

Michael D. McCool, Zheng Qin, and Tiberiu S. Popa. Shader metaprogramming. In *Proceedings of the ACM SIGGRAPH/EUROGRAPHICS conference on Graphics hardware*, HWWS '02, pages 57–68, Aire-la-Ville, Switzerland, Switzerland, 2002. Eurographics Association.

NVIDIA. *NVIDIA's Next Generation CUDA Compute Architecture: Kepler GK110 Whitepaper.* NVIDIA, Santa Clara, CA, USA.

NVIDIA. *CUDA Toolkit Documentation.* NVIDIA, Santa Clara, CA, USA, May 2013.

S. F. Reddaway. Dap–a distributed array processor. In *Proceedings of the 1st annual symposium on Computer architecture*, ISCA '73, pages 61–65, New York, NY, USA, 1973. ACM.

Chapter 3
Intel® Xeon Phi™ Coprocessors

Jim Jeffers

Abstract Intel recently launched the Intel® Xeon Phi™ Coprocessor to enhance the performance of the growing category of highly parallel scientific applications. This chapter provides an overview of the Intel Xeon Phi Coprocessor including its hardware and software architecture and the key usages that enable these highly parallel applications, like many in Geoscience to achieve new levels of performance while using familiar, standard programming models.

Keywords Intel® Xeon Phi™ coprocessor • Coprocessor MPI programming models

In late 2012, Intel launched its first in a new line of highly parallel "many-core" processing products, the Intel Xeon Phi Coprocessor. With up to 61 Intel Architecture general purpose processor cores plus a powerful, new wide vector processing unit in each core, these computationally rich processing engines target highly parallel, data intensive applications such as the GeoScience applications described in this book. In a world of rapidly growing power requirements in high performance data centers, an important benefit is the power efficiency (performance/watt) the coprocessor provides. One or more PCI Express compliant Intel Xeon Phi coprocessor cards can be added to enhance computational capabilities of an Intel® Xeon® processor platform or supercomputing cluster node.

Given that "peripheral attach" paradigm, one might immediately draw the conclusion that the coprocessor solely operates in a master-slave relationship with host Intel Xeon processor applications "offloading" computational tasks to the coprocessor occasionally. While this offload model is indeed robustly supported on the Intel Xeon Phi coprocessor, the hardware and software architecture enables a

J. Jeffers (✉)
Intel Corporation, 1800 River Rd, New Hope, PA 18938, USA
e-mail: james.l.jeffers@intel.com

X. Shi et al. (eds.), *Modern Accelerator Technologies for Geographic Information Science*,
DOI 10.1007/978-1-4614-8745-6_3, © Springer Science+Business Media New York 2013

broad set of existing programming models familiar to high performance computing application developers, system programmers and administrators.

For example, the Intel Xeon Phi coprocessor uses the same Intel development tools and product suites used to develop software for Intel Architecture processors, including compilers, performance libraries, and analysis tools. Even more illustrative is the fact that the coprocessor runs a standard Linux operating system, is IP addressable, supports well-known tools and protocols like the SSH, can mount file systems with NFS and all other standard Linux capabilities. In other words, the coprocessor can run complete self-contained applications and operates as a networked Linux-based computer.

In the remainder of this section, we will provide an overview of the Intel Xeon Phi coprocessor's hardware and software architecture that enables highly parallel applications to achieve extraordinary performance.

3.1 Hardware Architecture Overview

A symmetric multi-processor (SMP) on-a-chip is a good description for the Intel® Xeon Phi™ coprocessor. It provides up to 61 cores and significant reliability features while offering a familiar, well-known programming environment. When launched in November 2012, Intel Xeon Phi coprocessors were already in seven of the world's faster supercomputers (per the "Top 500" list, top500.org) and were used to build the world's most power-efficient supercomputer (per "Green 500" list, green500.org). The Intel Xeon Phi coprocessor is a true engineering marvel in many ways.

The ambitious goal for the coprocessor design was to simultaneously enable evolutionary *and* revolutionary paths forward for scientific discovery through efficient, high performance, technical computing. Evolutionary in creating a generally programmable solution that matches the training, application investments, standards, and computing environments of an existing, vibrant High Performance Computing (HPC) development community. Revolutionary in enabling powerful new parallel focused computing elements that give a new target for long term sustainable parallel programming optimization.

3.2 The Intel® Xeon Phi™ Coprocessor Family

The several different Intel Xeon Phi coprocessor models vary on such factors as performance, memory size and speed, thermal (cooling) solutions and form factor (type of card). All the coprocessor products interface to the Intel Xeon processor host platform through a PCI Express bus connection and consist of a coprocessor silicon chip on the coprocessor card with other components such as memory.

Figure 3.1 depicts the two types of double-wide PCI Express cards that are offered (passive and active cooling solutions). Passive heat sink cards will be used primarily

Fig. 3.1 Intel® Xeon Phi™ coprocessor card forms (top: passive cooling, bottom: active cooling)

in supercomputing cluster data centers where densely packed rack mount compute blades (nodes) will have high throughput cooling fans drawing air through the entire unit. Active fan sink cards will typically be used in desk-side workstation units.

3.3 Coprocessor Card Design

The coprocessor card can be thought of as a motherboard for a computer with up to 61 cores, complete with the silicon chip (containing the cores, caches and memory controllers), GDDR5 memory chips, flash memory, system management controller, miscellaneous electronics and connectors to attach into a computer system. A schematic view of the key components of the coprocessor card is shown in Fig. 3.2. The major computational functionality is provided by the Intel Xeon Phi coprocessor silicon chip. The silicon chip is, as you'd find with many Intel Xeon processors, contained in Ball Grid Array (BGA) packaging. This BGA package housing the silicon chip is the key component on a coprocessor card much like a processor is the key component on a computer motherboard.

Up to 16 channels of high bandwidth GDDR5 memory can be utilized and using a method known as *clamshell*, up to 32 memory devices can be attached using wire connections routed on both sides of the card, doubling the typical memory capacity.

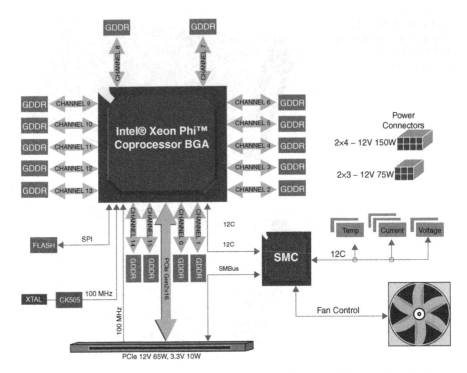

Fig. 3.2 Intel® Xeon Phi™ coprocessor card schematic. *Note*: On-board fan is only available on the 3100 series active product

Flash memory on the card is used to contain the coprocessor silicon's startup or bootstrap code, similar to the BIOS in an Intel Xeon processor platform.

The System Management Controller (SMC) handles monitoring and control chores such as: tracking card-level information from temperature, voltage, and current sensors, as well as adjusting the fan (if installed) accordingly to increase or decrease cooling capacity. The SMC provides the host's baseboard management controller (BMC) vital function status and control via the industry standard Intelligent Platform Management Interface (IPMI) over the System Management Bus (SMBus). The operating system software on the coprocessor chip communicates with the SMC via a standard I²C bus implementation.

3.4   Intel® Xeon Phi™ Coprocessor Silicon Overview

The Intel Xeon Phi coprocessor silicon implements the fundamental computational and I/O capabilities. As shown in Fig. 3.3, the many x86-based cores, the memory controllers, and PCI Express system I/O logic are interconnected with a high speed ring-based bidirectional on-die interconnect (ODI). Communication over the ODI is transparent to the running code with transactions managed solely by the hardware.

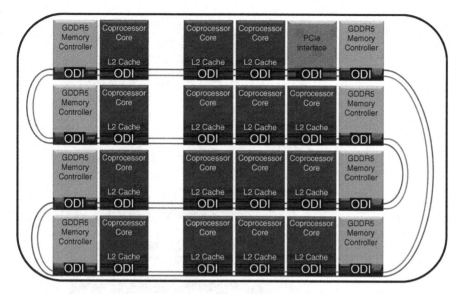

Fig. 3.3 Overview of the Intel® Xeon Phi™ coprocessor silicon and the on-die interconnect (ODI)

Each core has an associated 512-KB Level 2 (L2) cache to provide high speed, reusable data access. Furthermore, fast access to data in another core's cache over the ODI is provided to improve performance when the data already resides "on chip." Using a distributed Tag Directory (TD) mechanism, the cache accesses are kept "coherent" such that any cached data referenced remains consistent across all cores without software intervention.

From a software development and optimization perspective, a simplified way to view the coprocessor is as a symmetric multiprocessor (SMP) with a shared Uniform Memory Access (UMA) system; each core effectively having the same memory access characteristics and priority regardless of the physical location of the referenced memory.

3.5   Individual Coprocessor Core Architecture

A high level diagram of each processing core on the coprocessor silicon is shown in Fig. 3.4. The structure of the core implies the key design goals of creating a device optimized for high level of power-efficient parallelism while retaining the familiar, Intel architecture–based generally programmability. A 64-bit execution environment based on Intel64® Architecture is provided. Also, an in-order code execution model with round-robin multithreading is employed to reduce size, complexity, and power consumption of the silicon versus the deeply out-of-order, highly speculative execution support used primarily to improve serial-oriented code performance on

Fig. 3.4 Intel® Xeon Phi™
coprocessor individual core
structure

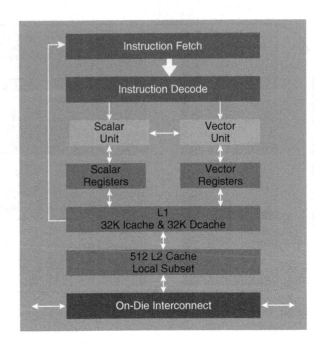

Intel Xeon processors. This difference in instruction processing flow also reflects how a programmer might consider partitioning and targeting code and applications for platforms that include Intel Xeon Phi coprocessors. The multithreading of the coprocessor plays a critical role in the in-order code execution model for achieving maximum performance and load balancing which makes it important for virtually all coprocessor applications to utilize. Coprocessor multithreading should not be confused with hyper-threading (HT) on Intel Xeon processors which uses opportunistic, event driven thread switching.

The core includes 32 KB each of L1 instruction (code) cache and L1 data cache, as well as the private (local) 512-KB L2 cache. Code is fetched from memory into the instruction cache and then goes through the instruction decoder for dispatch and execution. There are two primary instruction processing units. The scalar unit executes code using existing traditional x86 and x87 instructions and registers. The vector processing unit (VPU) executes the newly introduced Intel Initial Many Core Instructions (Intel® IMCI) utilizing a 512-bit wide vector length enabling very high computational throughput for both single precision and double precision calculations. Note that there is no support for MMX™ instructions, Intel Advanced Vector Extensions (Intel® AVX), or any of the Intel® Streaming SIMD Extensions (Intel® SSE). These instruction families were omitted to save space and power and to favor 512-bit SIMD capabilities unique to the Vector Processing Unit (VPU) of the Intel Xeon Phi coprocessor.

3.6 Coprocessor Software Architecture Overview

The unique thing for many first introduced to the Intel® Xeon Phi™ coprocessor operational software environment and architecture is its intentional goal to *not* be unique. As an Intel product, the coprocessor inherits many of the foundational elements and capabilities of widely used Intel Architecture processing platforms. These traits enable the coprocessor to be integrated as a peer processing platform into the well understood standard software infrastructure familiar to systems level developers, application developers and system administrators for both single workstation and large cluster supercomputers.

The most obvious indicator of broad capabilities is that the Intel Xeon Phi coprocessor bootstraps and runs a Linux operating system (OS) including Linux's substantial networking capability. The ability to run a Linux OS and communicate as a network peer is one of the primary reasons it is termed a *coprocessor* and not just an accelerator (which would require software applications to be managed by the host platform).

Another advantage is that the development tools such as compilers, profiling tools and debuggers can also be used in a similar manner as an Intel® Xeon® Processor. If fact, the Intel compilers, libraries and development tools suites are one in the same for both processors and coprocessor. So the Intel Xeon Phi Coprocessor development allows seamless use of standards and libraries such as OpenMP, Intel Math Kernel Library, Intel Thread Building Blocks, MPI and many others.

In this section, we look at the software architecture and components that enable the coprocessor to operate seamlessly in a standard environment. The architecture was designed to support a broad range of applications and programming models. Therefore, we will also discuss the variety of programming models available to take full advantage of the coprocessor capabilities.

There are two major components that comprise the software structure used to build and run applications and system services that utilize the coprocessor.

- *Development Tools and Runtimes*. The development tools and associated runtime services and libraries provided by tool packages such as Intel Parallel Studio XE 2013 and Intel Cluster Studio XE 2013. The Intel C/C++ compiler, Intel Fortran compiler, and Intel MPI library are some of the sub-components of these development tools packages.
- *Intel® Manycore Platform Software Stack (Intel® MPSS)*. The operational software specific to the coprocessor including middleware interfaces used by the development tools, device drivers for communication and control, coprocessor management utilities (such as the Control Panel), and the coprocessor's local Linux operating system. Collectively this set of software is known as the Intel Manycore Platform Software Stack.

When installing coprocessors for use in an Intel® Xeon® processor platform, an early step that normally needs to be done is to download, install, and launch the latest version of Intel MPSS (available at the intel.com/software/mic Web site). You

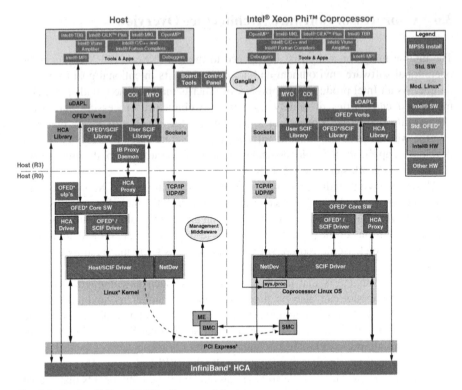

Fig. 3.5 The Intel® Xeon Phi™ software architecture

will then either access an existing toolset or install the available Intel or third party development tools including compilers.

Figure 3.5 shows a block diagram of the key components that comprise the coprocessor software architecture.

Figure 3.5 is broken into left and right sides as well as top and bottom halves, represented by the dashed lines. The left side corresponds to components on the host processor platform and the right side depicts software components on the coprocessor. In the next few sections, we will touch upon the key architectural concepts this diagram implies.

3.6.1 Symmetry

Looking at the left and right sides of Fig. 3.5 it should strike you that, with very few exceptions, the "boxes" on both sides of the diagram are the same. This belies a key underlying concept at the foundation of the software architecture for a platform with one or more coprocessors: that concept is symmetry. Making the coprocessor software components and interfaces symmetric with the host processor platform, in other words functionally identical, enables the coprocessor to be engaged by software developers in the same manner as they engage the processor.

This is a simple yet very powerful concept. This symmetry provides the fundamental basis supporting common development tools and common programming models on both the processor and coprocessor. Developers at all levels including those implementing tools, system software, and applications benefit by the familiar and common interfaces dramatically minimizing the learning curve and porting time for coprocessor targeted solutions of all kinds.

3.6.2 Ring Levels: User and Kernel

The top and bottom halves of the diagram in Fig. 3.5 represent the standard operating system notion of protection domain rings with user-level application code and system interface execution at ring 3 and more trusted, system level operating system kernel and driver code running at ring 0.

It is beyond our scope to more deeply discuss these operating system design areas other than to indicate access to devices and operating system kernel services occurs virtually always through counterpart ring 3 user mode library interfaces and kernel mode ring 0 modules. The user mode library interfaces manage the ring transitions between the companion kernel-level modules, maintaining the security and integrity of the system. The concept can be seen in Fig. 3.5 with like-named ring 3 user mode libraries calling upon their corresponding kernel module.

Now, we will discuss the programming models that are enabled and that influenced the creation of the coprocessor software architecture. Then the purpose of the provided system modules will have better context.

3.7 Coprocessor Programming Models and Options

The overall architecture of a platform that includes Intel Xeon Phi coprocessors enables a broad array of usages. This flexibility allows a dynamic range of solutions to address many target computing needs—from mostly serial processing to highly parallel processing to a mix of both. Intel and industry partners are delivering and creating tools and standards for processor/coprocessor platforms that can be used to develop applications that are optimal for the problem at hand.

Figure 3.6 illustrates the compute spectrum enabled when coupling processors and coprocessors. Depending on the application's compute needs, execution can be initiated on either a host processor or on one or more coprocessors. Depending on the application needs and system environment, any mix of computation between the processor and coprocessor can be chosen for optimal performance.

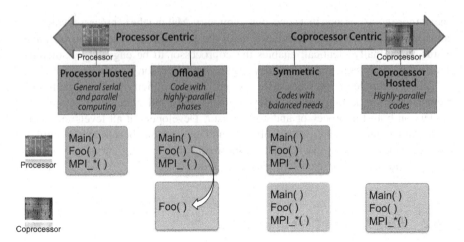

Fig. 3.6 The wide spectrum of joint programming models for an Intel® Xeon® platform with a coprocessor including primary application usages

Included in Fig. 3.6 is a conceptual view of how code might be launched and executed in the key enabled programming uses. From left to right those models are:

- *Processor Hosted.* The application is launched and executed on processors only.
- *Offload.* The application is launched and primarily managed on processors and selected portions of code (usually highly parallel) are run on coprocessors using either the Intel Math Kernel Library automatic offload capability, Intel Language Extensions for Offload or OpenMP "target" extensions.
- *Symmetric.* The application is launched on both coprocessors and processors with cooperative communication (typically via MPI).
- *Coprocessor Hosted.* The application is launched and executed on coprocessors only.

3.7.1 Breadth and Depth

The software architecture not only enables the breadth of models shown in Fig. 3.6 but also provides well known, deeper level fine controls similar to those available on other Intel Architecture processing platforms. Developers who want or need to focus on absolute maximum performance, are implementing targeted libraries and tools, or are developing specialty capabilities not otherwise fully enabled by existing tools are likely users of these programming alternatives. Figure 3.7 illustrates some of the layers and options available for developing applications with increasing levels of control, and, generally, more complexity.

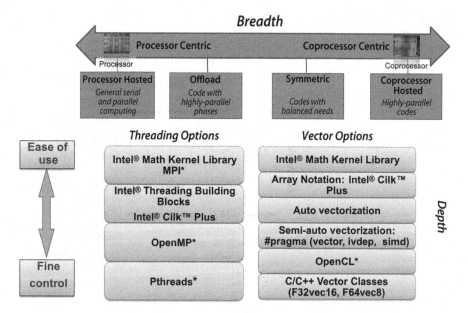

Fig. 3.7 Coprocessor programming options providing different methods of control for developers at all levels

3.7.2   Coprocessor MPI Programming Models

As previously mentioned, the coprocessor architecture has been designed to readily support the industry standard Message Passing Interface. In particular, the Intel® MPI library supports all the programming execution models described in Fig. 3.6. MPI is the de facto library-based communication environment used to enable parallel applications to run, communicate, and scale across multiple processing cores, either between the multiple cores in a single Intel Xeon processor platform or across a connected network of nodes (individual platforms) in a cluster. Furthermore, key MPI supporting sub-component standard layers such as the Open Fabrics Alliance defined Open Fabrics Enterprise Distribution (OFED) interfaces are also supported on both the processor and the coprocessor.

3.7.3   Offload Model

The offload model is characterized by the MPI communications taking place only between the host processors. The coprocessors are used exclusively through the offload capabilities provided as compiler extensions. This mode of operation is fairly straightforward, simply enabling offloading as part of existing or newly developed MPI-based applications. Making MPI library calls inside offloaded code is not supported. Figure 3.8 illustrates the MPI with offload model.

Fig. 3.8 MPI on the host
processor platform using
offload to coprocessors

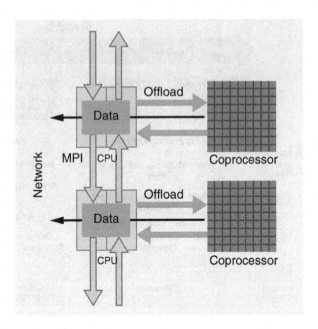

Fig. 3.8 MPI on the host processor platform using offload to coprocessors

3.7.4   Coprocessor-Only Model

The MPI coprocessor-only or *native* model has the MPI processes launched and residing solely on the coprocessor. MPI libraries, the application, and other needed libraries are uploaded to the coprocessors. Then an application can be launched from the host or from the coprocessor. Once the application is running, MPI network communications between other coprocessors (either on the local node or to other network fabric connect nodes) are managed by the Intel® Coprocessor Communications Link (Intel® CCL) services. Intel CCL provides underlying services to the MPI library to select the optimal transport for MPI messages. One such Intel CCL transport mechanism is the peer-to-peer PCI Express DMA support to directly transfer message data between the coprocessor's memory and a peer InfiniBand (IB) adapter without host memory staging. Figure 3.9 illustrates the MPI Coprocessor-Only model.

3.7.5   Symmetric Model

The MPI Symmetric programming model launches and executes the MPI application on both the host processor and the coprocessors. Figure 3.10 illustrates the symmetric MPI model. This is the most flexible model supporting "any to any" messaging. Message passing may occur within the coprocessor, within the host processor, between the coprocessor and the processor within the same node, and between coprocessors and processors across a cluster through several fabric

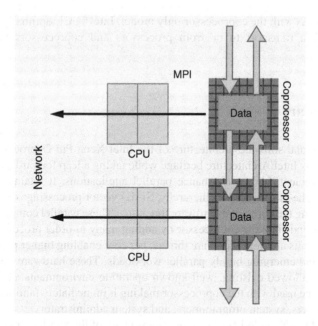

Fig. 3.9 MPI running on coprocessors only

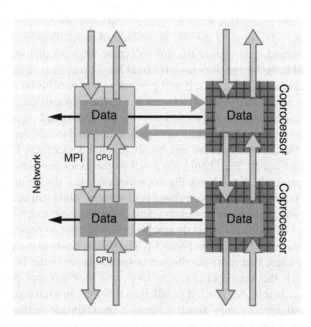

Fig. 3.10 MPI symmetric communications with MPI running on both processors and coprocessors

mechanisms. As with the coprocessor only model, Intel® CCL optimally manages communication transport to or from processors and coprocessors across the network.

3.8 Summary

The hardware and software architecture of the Intel Xeon Phi Coprocessor draws heavily from its Intel Architecture heritage while taking a leap forward as a vehicle for energy efficient, high performance parallel applications. It retains a familiar multi-level cache-based memory hierarchy, SIMD vector processing, shared memory usage among the cores, and hardware threading. These parallel computing capabilities are extended on the coprocessor by adding many in-order processing cores, wider vector units and four hardware threads per core enabling higher performance for existing and emerging highly parallel workloads. These hardware architecture features have allowed existing, well-known operating environments and development tools to be used with the coprocessor making it immediately familiar to application developers, system programmers, and system administrators.

Since the Intel Xeon Phi coprocessor retains many of the familiar characteristics of the Intel Xeon processor, it is useful to understand how to compare their parallel computing capability. Advice for successful parallel programming can be summarized as "Program with lots of threads that use vectors with your preferred programming languages and parallelism models." Since most applications have not yet been structured to take advantage of the full magnitude of parallelism available in any processor, understanding how to restructure to expose more parallelism is critically important to enable the best performance for Intel Xeon processors or Intel Xeon Phi coprocessors. This restructuring itself will generally yield benefits on most modern general-purpose computing systems, a bonus due to the emphasis on common programming languages, models, and tools across the processors and coprocessors.

It has been said that a single picture can speak a thousand words; for understanding Intel Xeon Phi coprocessors (or any highly parallel device) it is Fig. 3.11 that speaks a thousand words. You should not dwell on the exact numbers. The picture speaks to this principle: Intel Xeon Phi coprocessors offer the ability to make a system that can potentially offer exceptional performance while still being buildable and power efficient. Intel Xeon processors deliver performance much more readily for a broad range of applications but do reach a practical limit on peak performance as indicated by the end of the line in Fig. 3.11. The key is "ready to use parallelism." Note from the picture that more parallelism is needed to make the Intel Xeon Phi coprocessor reach the same performance level, and that requires programming adapted to deliver that higher level of parallelism required. In exchange for the programming investment, we may reach otherwise unobtainable performance. The advantage of these Intel products is that the use of the same parallelism model, programming languages, and familiar tools to greatly enhance preservation of programming investments.

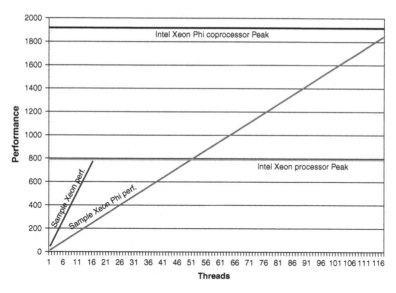

Fig. 3.11 This picture speaks a thousand words

3.9   For More Information

Some additional reading worth considering includes:

- "Intel® Xeon Phi™ Coprocessor High Performance Programming",
  Jim Jeffers, James Reinders, © 2013, publisher: Morgan Kaufmann. http://lotsofcores.com
- "Intel® Xeon Phi™ Coprocessor High Performance Programming"
- "An Overview of Programming for Intel® Xeon® processors and Intel® Xeon Phi™ coprocessors" (Intel 2012) available at http://tinyurl.com/xeonphisum
- Intel® Xeon Phi™ Coprocessor: Intel Developer Zone, documentation and additional information is also available at http://intel.com/software/mic

Chapter 4
Accelerating Geocomputation with Cloud Computing

Qunying Huang, Zhenlong Li, Jizhe Xia, Yunfeng Jiang,
Chen Xu, Kai Liu, Manzhu Yu, and Chaowei Yang

Abstract The scientific and engineering advancements in the twenty-first century pose computing intensive challenges in managing Big Data, using complex algorithms to extract information and knowledge from Big Data, and simulating physical and social phenomena. Cloud computing is considered as the next generation computing platform with the potential to address these computing challenges and redefine the possibilities of geoscience and digital Earth. This chapter introduces through examples how cloud computing can help accelerate geocomputation with: (1) easy and fast access to computing resources that can be available in seconds to minutes, (2) elastic computing resources to handle spike computing loads, (3) high-end computing capacity to address large-scale computing demands, and (4) distributed services and computing to handle the distributed geoscience problems, data and users.

Keywords Spatial computing • Geosciences • Field-programmable gate array (FPGA) • Many integrated core (MIC) • Parallel computing • CyberGIS

Q. Huang
Department of Geography, University of Wisconsin,
Madison, WI 53706, USA

Z. Li • J. Xia • Y. Jiang • C. Xu • K. Liu • M. Yu
Center for Intelligent Spatial Computing, George Mason University,
Fairfax, VA 22030-4444, USA

C. Yang (✉)
Department of Geography and GeoInformation Science, Center for Intelligent Spatial
Computing, George Mason University, 4400 Univ. Dr., Fairfax, VA 22030-4444, USA
e-mail: cyang3@gmu.edu

X. Shi et al. (eds.), *Modern Accelerator Technologies for Geographic Information Science*,   41
DOI 10.1007/978-1-4614-8745-6_4, © Springer Science+Business Media New York 2013

4.1   Introduction

Petabytes of geoscience data are collected on a daily base. Mining information and knowledge from the data and using the data to simulate large-scale phenomena call for a computing infrastructure to efficiently process these computing tasks (Yang et al. 2011a). However, most traditional computing infrastructure lacks the agility to keep up with the accelerating demands for more computing resource. Cloud computing, a new distributed computing paradigm, can quickly provision computing resource in an on-demand fashion. This can be utilized to address Geoscience challenges of computing, data and concurrent intensities (Yang et al. 2011b). This chapter introduces how cloud computing can help accelerate geocomputation in four aspects through examples:

- *Computing availability.* A typical timespan for procuring traditional computing infrastructure is weeks to months and significant human resources are spent. After the procurement of the infrastructure, more effort for administrative tasks is required for daily maintenance. Cloud computing has demonstrated the capacity to accelerate the procurement of computing resource and reduce dedicated administrative cost (Yang et al. 2011b).
- *Computing elasticity.* With the availabilities of on-demand computing resources (e.g., computing power, storage and networking), cloud computing can help applications to handle spike computing requirement without long-term commitment (Huang et al. 2013b). For example, Amazon Elastic Cloud Computing (EC2) provides auto-scaling service, allowing cloud consumers to scale Amazon EC2 computing capacity up or down automatically according to pre-defined conditions (e.g., central processing unit [CPU] utilization, and the user concurrent access number).
- *High performance computing (HPC) capability.* Infrastructure as a Service (IaaS), a category of popular cloud services, can easily offer high-end computing capabilities. For example, Amazon EC2 Cluster, with 17024 CPU cores in total, a clock speed of 2.93 GHz per core, and 10G Ethernet network connection, was ranked as 102*th* on the TOP 500 supercomputer lists in November 2012.[1] The HPC capability of cloud computing can be easily leveraged to support critical scientific computing demands (Huang et al. 2013b; Rehr et al. 2010).
- *Service and computing distribution.* Cloud computing natively supports processing of distributed data, problems, and users (Yang et al. 2011b). On one hand, cloud computing providers offer computing and storage services that are globally distributed (for example, three major cloud providers, Amazon, Microsoft and Google, have multiple data centers around the world). On the other hand, the geoscience data, users, and problems are globally dynamically distributed.

The following sections introduce how cloud computing can help accelerate geocomputation from these four aspects in details.

---

[1] http://www.top500.org/system/177457

4.2 Computing Availability

Better access to computing resources is one of the reasons that cloud users are transiting from traditional computing to cloud computing. Comparing to using traditional computing, leveraging cloud computing can help users reduce the time spent on preparing infrastructure (e.g., purchasing physical machines, and configuring the networking). In addition, system administration and maintenance works are significantly reduced (Huang et al. 2013b). Table 4.1 compares the average time spent on deploying and operating dust storm model on Amazon EC2 to that of a local cluster.

Only several hours to several days are required for deploying new cloud computing virtual machines (VMs) for model simulations based on either using public Amazon machine image (AMI) or hardening image from scratch. Using public available AMI other than hardening the image from scratch could leverage preconfigured basic operating system (OS) package and required cluster software packages, such as the scheduling software MPICH2.[2] This could help, for example the dust storm model, reduce the configuration effort significantly from several days to several hours. After the first deployment, EC2 users can use the deployed image to launch as much as VMs as needed in a few minutes to form a cloud based cluster (Table 4.1). Considering the time of purchasing the servers and configuring the hardware and software, a traditional HPC cluster requires at least several weeks to set up the deployment environments for the first time.

In addition to using private physical HPC environments, users can also apply public HPC computing resources. However, the approval process may take several weeks to even months. For example, requests for accessing XSEDE, a single virtual

Table 4.1 Average time spent on deploying dust storm model onto Amazon EC2 and local cluster (revised based on Huang et al. 2013b)

Items	Local cluster	Amazon EC2	Options for cloud environment
Procure cluster	~4 weeks	None	N/A
Configure cluster operating system (OS)	~1 week	None	Use a public AMI with OS installed
		~1 week	Harden image from scratch
Configure dust storm model	~1 day	~2 h	Use a public AMI with most required software dependencies installed
		~1 days	Harden image from scratch
Start cluster	120 s	45 s	N/A
Stop cluster	60 s	57 s	
Resume cluster	N/A	45 s	
Total time needed for the first time deployment	~5 weeks	~2 h	Use a public AMI
		~1 week	Harden image from scratch

---

[2] http://phase.hpcc.jp/mirrors/mpi/mpich2/

system that scientists can use to interactively share computing resources, are reviewed quarterly by the XSEDE Resource Allocation Committee (XRAC) and users should summit the requests 2.5 months in advance.[3]

Comparing to both building up a private HPC or applying a public HPC account, cloud computing can accelerate geocomputation by getting access to the required computing resources in a few minutes instead of weeks or months.

4.3 Computing Elasticity

One of the most important characteristics of cloud computing is elasticity (Mell and Grance 2011). With elasticity, applications, running on the cloud, can increase the amount of computing resources to handle spike workloads and accelerate geocomputation in a few seconds to minutes. Computing resources can be released once the workloads decreased. In geoscience applications, elasticity is critical since they may require the allocation of dynamic computing resources dynamically. For example, responding to natural disasters (such as, earthquakes, wildfires and tsunami) requires elastically bringing up more computing resources to handle the spike requests from the public and decision makers (Huang et al. 2013b; Yang et al. 2011b).

This section uses Global Earth Observation System of Systems (GEOSS) Clearinghouse (CLH) as an example to demonstrate how to use elasticity to handle spiking workloads. GEOSS CLH is the engine of GEOSS common infrastructure to provide the capabilities of managing, publishing, harvesting, and searching metadata (Liu et al. 2011). As a global operational system, GEOSS CLH has been intensively accessed by general end users, developers and other Geospatial Cyberinfrastructure (GCI) communities (e.g. Group on Earth Observations [GEO] portal). From December 22nd 2010 to October 31st 2012, the total number user access is 2,202,660. By analyzing the GEOSS CLH log file, two user access patterns that result spike workloads are discovered:

- United States and Europe have a large number of end users who generate massive concurrent accesses.
- The user access frequency increases and decreases daily with spike workloads in specific time periods (normally at local morning and afternoon hours).

The concurrent intensity can cause performance issues such as slow response, network congestion and even system failure of the GEOSS CLH operation. To address these issues, the elasticity of cloud computing is adopted.

Figure 4.1 shows the performance of GEOSS CLH on Amazon EC2 (average response time measured in seconds) by using elasticity to handle different numbers of concurrent requests. The x-axis indicates the concurrent access number and the y-axis indicates the average response time. Utilizing elasticity in Amazon EC2

---

[3] https://www.xsede.org/web/guest/allocation-policies#uses:eligibility

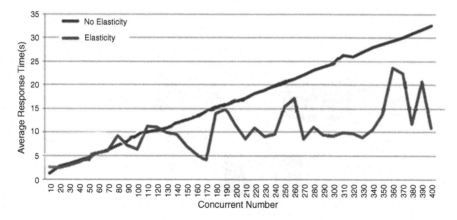

Fig. 4.1 EC2 response performance comparisons by single and five autoscaling instances

requires predefined elasticity rules. For instance, the rule could be defined as: when the user request number exceeds a certain limit, or the latency time is longer than certain seconds (4 seconds in this experiment), a new GEOSS CLH instance will be automatically launched.[4] In addition, the maximum number of GEOSS CLH instances can be predefined (five instances in this experiment).

According to the results, elasticity can accelerate the geocomputation and reduce the average response time. When GEOSS CLH has received 50–60 concurrent requests, where average response time is longer than 4 seconds, a new GEOSS CLH instance was launched according to the predefined elasticity rules. However, the new instance cannot respond user quests instantly and require around 1 minute to boot up. This causes the average time started to drop until the concurrent number reaches 80. GEOSS CLH started to launch the third, fourth or fifth instance when the concurrent access number is 170, 210 or 270. In addition, when workloads are reduced to a certain level, additional GEOSS CLH instances will be terminated to release computing resources.

4.4 HPC Capability

Numerical methods and complex algorithms are computing intensive. And performing complex computing tasks requires the availability of a large number of computing resources, and serial computing using a single computer is not sufficient any more (Huang et al. 2013a). Traditionally, these needs have been addressed by using HPC facilities such as clusters and supercomputers (Huang and Yang 2011), which are difficult to configure, maintain and operate (Vecchiola et al. 2009). Therefore, geoscientists are seeking for high-end computing technologies with better flexibility and configurability.

---

[4] http://aws.amazon.com/autoscaling/

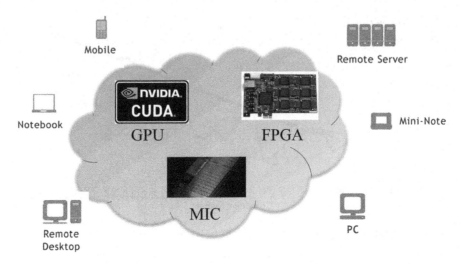

Fig. 4.2 Cloud computing platforms can provide the latest high end computing devices

Cloud computing provides scientists with a complete new computing paradigm for accessing and utilizing the computing infrastructure. Cloud computing services, especially IaaS, can be easily adopted to offer the prevalent high-end computing technologies to provide more powerful computing capabilities (Fig. 4.2). A range of diverse computing resources for users' computing needs can be offered, such as Many Integrated Cores (MICs), Graphics Processing Units (GPUs), and Field Programmable Gate Arrays (FPGAs). For example, Amazon provides GPU-based VMs in a scalable, elastic environment on a fast, low latency and non-blocking 10 Gbps network.

- *GPU Computing*
  GPU computing technology becomes popular in the past years (Nickolls and Dally 2010). As a specialized circuit, GPU is initially designed to accelerate image processing (Pharr and Fernando 2005). With the distinguished parallel computing capabilities for processing large volume data, GPU computing began to be used in geocomputation in the past several years. For example, GPUs have been demonstrated to be as excellent computing resources to address geoscience problems such as computational fluid dynamics, and seismic wave propagation (Walsh et al. 2009). Li et al. (2013) have utilized many-core GPUs to improve the performance of visualizing 3D/4D environmental data. High performance GPUs also enables accelerating batch processing of spatial raster data by performing hundreds of arithmetical operations in parallel (Steinbach and Hemmerling 2012).

  Currently, GPU computing has been gradually integrated into cloud computing infrastructure, which is called GPU-based cloud computing (GCC). On one hand, GCC provides an additional option for applications that can benefit from the efficiency of the parallel computing capability of GPUs.[5] On the other hand, GCC can provide more benefits for applications by utilizing cloud capabilities (e.g., fast accessibility in Sect. 4.2, and elasticity in Sect. 4.3) compared to that of leveraging GPU outside of cloud computing platforms. Therefore, GCC is now used in a variety of scientific research and applications. For example, Wang and Shen (2011) used GCC to support intelligent transportation management. Sugumaran et al. (2011) compared the performance of CPU and GPU on cloud platform for processing massive Light Detecting and Ranging (LiDAR) topographic data.

- *MIC Computing*
  Inter® MIC Architecture combines 50 or more Inter® CPU cores into a single chip.[6] The MIC architecture supports a high degree of parallelism with smaller, lower-power performance Intel® processor core. For example, the microprocessor, "Single-Chip Cloud Computer" (SCC), created by the Intel Labs, is capable of scaling to 100 cores and beyond.[7] Because of its high degree of parallelism, MIC architecture has been used by many scientists to solve scientific problems. For example, Wald (2012) investigated how to efficiently build bounding volume hierarchies with surface area heuristic on the MIC architecture. Satish et al. (2010) have presented a competitive analysis of comparison and non-comparison based sorting algorithm on CPUs and GPUs, and implemented and tested the algorithm on the MIC architectures.

  The ever increasing HPC requirement needs more research and developments for new computer architectures. As a new multiple processor computer architecture, MIC architecture should be integrated into cloud computing to empower high-end parallel computing capability.

- *FPGA Computing*
  FPGA is an integrated circuit that contains a large number of (64 to over 10,000) identical and programmable logic blocks, and a programmable routing which allows the logic block inputs and outputs to be connected to form larger circuits (Betz and Rose 1999). Field Programmable (FP) means that the FPGA's function is defined by a user's program rather than by the manufacturer of the device.[8] FPGAs show very high computing capability for image processing, with high degree of parallelism and a large number of internal memory banks (Saegusa et al. 2008). Underwood (2004) demonstrated that at one point an FPGA had a higher peak floating-point performance than a CPU core in simple operations such as single and double precision compliant floating-point addition, multiplication and division. This is because FPGAs are able to customize the allocation of resources to meet the needs of the application while CPUs have fixed functional units.

---

[5] http://aws.amazon.com/gpu/

[6] www.intel.com

[7] http://www.intel.com/content/www/us/en/research/intel-labs-single-chip-cloud-computer.html

[8] http://www.epanorama.net/links/fpga.html

In addition to good computing performance, FPGAs also present interesting opportunities for fault tolerance due to their ability to be reconfigured (Stott et al. 2008). Therefore, it is possible to provide more reliable cloud computing services using FPGAs-based computing system than that of traditional CPU-based system. Some creative research have been done on FPGAs in cloud computing architecture. For example, Shu et al. (2012) presented a DaaS (Desktop-as-a-Service) with cloud server technologies on FPGA to address the problem caused by high power consumption and heavy network traffic.

4.5 Service and Computing Location Distribution

Cloud computing provides distributed computing resources to support distributed data processing and computing requirements. Cloud computing could accelerate distributed geospatial computation in at least two aspects: (1) easily integrating cloud computing platforms from different organizations to build a larger computing pool; and (2) optimizing geospatial applications by considering spatiotemporal distributions of data, computing resources and users.

- *Distributed computing resource integration*
  Geoscience applications are often data and compute intensive due to the underlying large volume, complex and high dimensional spatiotemporal datasets to be processed, analyzed and visualized (Yang et al. 2011b). Users of these applications require the results to be returned in a timely manner or even real time. For example, dust storm modeling is a typical compute intensive process and near real time dust storm simulation is essential for dust storm forecasting. Single computer, even with high-end hardware configurations, is far from enough to fulfill this requirement. Cloud computing offers the possibility to build a distributed computing infrastructure by leveraging less expensive commodity computers all over the world. Organization can contribute distributed resources to form a larger computing pool. This distributed infrastructure can serve as a spatial cloud computation (SCC) platform for best utilizing, managing and supporting big data processing, accessing, discovery, and dissemination. Building such a SCC platform is better than that of traditional distributed computing paradigm in that (1) the virtualization technology can be used to encapsulate the underlying computing infrastructure to make it more usable; (2) cloud computing provides the capabilities of the elasticity for different geoscience applications (Yang et al. 2011b; Sect. 4.3).
- *Geospatial application deployment optimization*
  Deploying geoscience applications onto cloud computing enables users to take the advantage of elasticity, scalability and high-end computing capabilities offered by cloud computing. However, in order to optimize the deployment, spatiotemporal patterns of computing resources, storage resources and users need to be considered. These spatiotemporal patterns can be mined from the following aspects (1) the physical location of computing and storage resources, (2) the

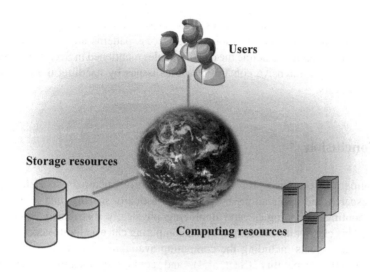

Fig. 4.3 Relationship among users, computing resources and geospatial resources

distribution of potential users, and (3) the dynamic massive concurrent access for users at different locations. Figure 4.3 depicts the relationship among users, computing resources and storage resources in a distributed cloud-based environment.

The first aspect to be considered is the spatial distribution of computing and storage resources. When deploying applications onto cloud services, the virtualization capability of cloud computing makes the unified computing and storage resources available at all cloud regions. However, the actual computing resources (e.g., underlying physical computing resources of a VM) and storage resources (e.g., data centers that store geospatial datasets) are physically distributed. For example, Amazon has multiple global data centers and edge locations for applications to better deliver content to global end users with lower latency.[9] Examining the distribution pattern of the underlying physical resources as well as considering the application characteristics will help us better deploy the application and choose where to store geospatial data to improve the overall system performance.

Another aspect is the spatial distribution of users and their temporal access patterns of the application. For example, if an application is mostly accessed during daytime, then users from North America and Asia are less likely to access the application at the same time due to the time difference of the two continents. Exploring the users' spatiotemporal patterns is helpful to address concurrent issues when many users are interacting with an application simultaneously (Yang et al. 2011b). This is especially important for the concurrent intensive applications,

---

[9] http://aws.amazon.com/cloudfront/

since intensive concurrent accesses may cause serious performance issues (Sect. 4.3). Once the users' spatiotemporal access patterns are identified, these patterns can be applied to define the elasticity rules mentioned in Sect. 4.3 to effectively address the intensive concurrent access issues by handling the load with scalable servers.

4.6 Conclusion

Cloud computing becomes a promising computing infrastructure to accelerate geoscience research and applications by pooling, elastically sharing, and integrating latest computing technologies, and deploying physically distributed computing resources. This chapter discusses how cloud computing can accelerate geocomputation from four aspects including the computing availability (Sect. 4.2), elasticity (Sect. 4.3), HPC capability (Sect. 4.4), and computing location distribution (Sect. 4.5). The benefits brought by cloud computing to accelerate geocomputation are also helping redefine the possibility of geosciences and digital Earth for the twenty-first century (Yang et al. 2013).

References

Betz V, Rose J FPGA routing architecture: Segmentation and buffering to optimize speed and density. In: Proceedings of the 1999 ACM/SIGDA seventh international symposium on Field programmable gate arrays, 1999. ACM, pp 59–68.

Huang Q, Yang C (2011) Optimizing grid computing configuration and scheduling for geospatial analysis: An example with interpolating DEM. Computers & Geosciences 37(2):165–176.

Huang Q, Yang C, Benedict K, Rezgui A, Xie J, Xia J, Chen S (2013a) Using adaptively coupled models and high-performance computing for enabling the computability of dust storm forecasting. International Journal of Geographical Information Science 27(4):765–784.

Huang Q, Yang C, Benedict K, Chen S, Rezgui A, Xie J (2013b) Utilize cloud computing to support dust storm forecasting. International Journal of Digital Earth 6(4):338–355.

Li J, Jiang Y, Yang C, Huang Q, Rice M (2013) Visualizing 3D/4D environmental data using many-core graphics processing units (GPUs) and multi-core central processing units (CPUs). Computer & Geosciences 59:78–89.

Liu K, Yang C, Li W, Li Z, Wu H, Rezgui A, Xia J The GEOSS Clearinghouse high performance search engine. In: Geoinformatics, 2011 19th International Conference on, 2011. IEEE, pp 1–4.

Mell P, Grance T (2011) The NIST definition of cloud computing (draft). NIST special publication 800:145.

Nickolls J, Dally WJ (2010) The GPU computing era. Micro, IEEE 30(2):56–69.

Pharr M, Fernando R (2005) Gpu gems 2: programming techniques for high-performance graphics and general-purpose computation. Addison-Wesley Professional ©2005.

Rehr JJ, Vila FD, Gardner JP, Svec L, Prange M (2010) Scientific computing in the cloud. Computing in Science & Engineering 12(3):34–43.

Saegusa T, Maruyama T, Yamaguchi Y How fast is an FPGA in image processing? In: Field Programmable Logic and Applications, 2008. FPL 2008. International Conference on, 2008. IEEE, pp 77–82.

Satish N, Kim C, Chhugani J, Nguyen AD, Lee VW, Kim D, Dubey P (2010) Fast sort on cpus, gpus and intel mic architectures. Technical report, Intel.

Shu S, Shen X, Zhu Y, Huang T, Yan S, Li S Prototyping Efficient Desktop-as-a-Service for FPGA Based Cloud Computing Architecture. In: Cloud Computing (CLOUD), 2012 IEEE 5th International Conference on, 2012. IEEE, pp 702–709.

Steinbach M, Hemmerling R (2012) Accelerating batch processing of spatial raster analysis using GPU. Computers & Geosciences 45:212–220.

Stott E, Sedcole P, Cheung P Fault tolerant methods for reliability in FPGAs. In: Field Programmable Logic and Applications, 2008. FPL 2008. International Conference on, 2008. IEEE, pp 415–420.

Sugumaran R, Oryspayev D, Gray P GPU-based cloud performance for LiDAR data processing. In: Proceedings of the 2nd International Conference on Computing for Geospatial Research & Applications, 2011. ACM, p 48.

Underwood K FPGAs vs. CPUs: trends in peak floating-point performance. In: International Symposium on Field Programmable Gate Arrays: Proceedings of the 2004 ACM/SIGDA 12 th international symposium on Field programmable gate arrays, 2004. vol 24. pp 171–180.

Vecchiola C, Pandey S, Buyya R High-performance cloud computing: A view of scientific applications. In: Pervasive Systems, Algorithms, and Networks (ISPAN), 2009 10th International Symposium on, 2009. IEEE, pp 4–16.

Wald I (2012) Fast Construction of SAH BVHs on the Intel Many Integrated Core (MIC) Architecture. Visualization and Computer Graphics, IEEE Transactions on 18(1):47–57.

Walsh SD, Saar MO, Bailey P, Lilja DJ (2009) Accelerating geoscience and engineering system simulations on graphics hardware. Computers & Geosciences 35(12):2353–2364.

Wang K, Shen Z (2011) Artificial societies and GPU-based cloud computing for intelligent transportation management. Intelligent Systems, IEEE 26(4):22–28.

Yang C, Goodchild M, Huang Q, Nebert D, Raskin R, Xu Y, Bambacus M, Fay D (2011a) Spatial cloud computing: how can the geospatial sciences use and help shape cloud computing? International Journal of Digital Earth 4(4):305–329.

Yang C, Wu H, Huang Q, Li Z, Li J (2011b) Utilizing spatial principles to optimize distributed computing for enabling physical science discoveries. Proceedings of National Academy of Sciences 108(14):5498–5503.

Yang C, Xu Y, Nebert D (2013) Redefining the Possibility of Digital Earth and Geosciences with Spatial Cloud Computing, International Journal of Digital Earth 6(4):1–8.

Part III
MAT in GIScience Applications

Chapter 5
Parallel Primitives-Based Spatial Join of Geospatial Data on GPGPUs

Jianting Zhang

Abstract Modern GPU architectures closely resemble supercomputers. Commodity GPUs that have already been integrated with personal and cluster computers can be used to boost the performance of spatial databases and GIS. In this study, we report our preliminary work on designing and implementing a spatial join algorithm on GPUs by using generic parallel primitives that have been well understood and efficiently implemented in many parallel libraries. In addition to help understand the inherent data parallelisms in spatial join operations, our experiments have shown that the reference implementation, which represents a tradeoff between code efficiency and code complexity, is able to achieve a 6.7× speedup when compared to an optimized CPU serial implementation. The result is encouraging in the sense that native implementation of spatial joins directly on top of GPU accelerators can potentially achieve much higher speedups for spatial joins which are fundamental to spatial databases and vector GIS. The implementations of parallel spatial algorithms on top of generic parallel primitives can be an important first step towards designing and developing high-performance spatial-specific parallel primitives to make it easier to build parallel spatial databases and GIS.

Keywords Spatial join • GPU • High-performance • Parallel primitives

5.1 Introduction

Spatial joins are fundamental in Spatial Databases and Geographical Information System (GIS). Given two geospatial datasets (which can be points, polylines and polygons), a spatial join finds all pairs of objects satisfying a given spatial

J. Zhang (✉)
Department of Computer Science, The City College of New York,
138 Street at Convent Avenue, NAC 8/206, New York, NY 10031, USA
e-mail: jzhang@cs.ccny.cuny.edu

X. Shi et al. (eds.), *Modern Accelerator Technologies for Geographic Information Science*, 55
DOI 10.1007/978-1-4614-8745-6_5, © Springer Science+Business Media New York 2013

relationship between the objects, such as within, intersect and nearest neighbor. Spatial joins on CPUs have been extensively studied over the past few decades (Jacox and Samet 2007) given their practical importance. However, while research in parallel spatial joins can be dated back to 1990s (Hoel and Samet 1994), it was not until General Computing on Graphics Processing Units (GPGPU)[1] technologies on commodity hardware become available in recent years that using parallel spatial join processing to speed up Spatial Databases and GIS performance starts to be practical, both technologically and economically. As argued in Clematis et al. (2003), despite significant research on parallel geospatial processing, research before 2000 has very little impact on real practices at large due to quite a few factors, especially limited accesses to parallel hardware. On the other hand, the current GPU architectures closely resemble supercomputers as both implement the primary Parallel Random Access Machine (PRAM) characteristic of utilizing a very large number of threads with uniform memory latency (Hong et al. 2011). As such, we believe research on efficient spatial joins on GPGPUs is timely and can potentially have a large impact on the geospatial computing community.

It is well known that spatial joins have two phases, i.e., filter and refinement (Jacox and Samet 2007). The optional filter phase relies on various spatial indexing data structures to filter out a large portion of candidate pairs to be joined while the refinement phase computes spatial relationships among filtered candidate pairs. A few GPGPU-based spatial indexing data structures have been proposed in the past few years (Zhou et al. 2008, 2011; Zhang et al. 2010; Hou et al. 2011; Luo et al. 2011) and can be applied in the filter phase. In this study, we will be focusing on the refinement phase. Since many spatial operations that are involved in the refinement phase, such as calculating distances and point-in-polygon tests, are more computing intensive than testing the spatial relationships based on Minimum Bounding Boxes (MBB) in the filter phase, it is more desirable to use GPUs to speed up the refinement phase.

While computing efficiency is the driving motivation for GPGPU-based parallel spatial joins, we believe it is also important to understand the inherent parallelisms in spatial joins so that the proposed parallel spatial join algorithms can sustain across different GPU architecture generations and interoperate with multi-core CPUs which increasingly have more GPU features as the numbers of CPU cores increase. As such, instead of directly providing a CUDA-based implementation, our prototypical implementation adopts a parallel primitive-based approach and the implementation is based on Thrust library[2] that comes with CUDA distributions since version 4.0.[3] The implementation can serve as a baseline to compare with both a serial CPU implementation and a native implementation that uses CUDA directly which is under development. It is well-known that parallel primitives based implementations represent tradeoffs between coding complexity and code efficiency. Spatial join on multi-dimensional geographical data on top of generic parallel primitives for one-dimensional vectors may not be the most efficient ones.

---

[1] http://en.wikipedia.org/wiki/GPGPU

[2] http://code.google.com/p/thrust/

[3] http://developer.nvidia.com/cuda-toolkit-40

Nevertheless, the prototypical implementation can serve as a starting point for developing efficient spatial-specific parallel primitives to make it easier to build parallel Spatial Databases and GIS.

Our preliminary results show that, the reference implementation is able to achieve a 6.7× speedup when compared with an optimized serial CPU implementation when joining pickup locations of 122,043 taxi trip records (points) with 43,252 MapPluto tax lots[4] (polygons) that have 293,335 vertices in the New York City (NYC) area. We expect significant higher speedups can be achieved when the spatial join algorithm is implemented in CUDA and provided as a parallel geospatial computing primitive. In addition, the reference implementation is more than 256× faster than joining the same two datasets using the state-of-the-art open source geospatial packages for indexing (libspatialindex[5]) and distance computation (GDAL/OGR[6]). The result clearly shows the potential of the performance gains in evolving existing Spatial Databases and GIS software that are optimized for traditional hardware architectures to modern hardware architectures. The features we have exploited in this study include simple array-based cache conscious data structures (vs. sophisticated pointer-based data structures using dynamic memory allocations), in-memory processing (vs. disk-resident) and parallel accelerations (vs. using standalone uni-processors).

The rest of the paper is arranged as follows. Section 5.2 introduces the background and formulates the research problem. Section 5.3 presents the design of the proposed spatial join algorithm using parallel primitives. Section 5.4 provides implementation details and experiment results, including the comparisons with a baseline serial CPU implementation and an open source implementation using existing GIS software stack. Finally Sect. 5.5 concludes and discusses future research directions.

5.2 Background and Problem Formulation

Given two vector geospatial datasets, which can be point, polyline or polygon data types, spatially joining the two datasets can be performed when neither of the datasets, either of the datasets or both of the datasets are indexed (Jacox and Samet 2007). While traditionally indexing geospatial data is considered expensive and spatial join techniques have been developed for non-indexed spatial data, recent works have shown that spatial indexing can be efficiently performed on GPUs (Zhou et al. 2008, 2011; Zhang et al. 2010; Hou et al. 2011; Luo et al. 2011). Furthermore, many geospatial data can be considered static relative to their update cycles and spatial indexing can be done offline. As such, we assume both datasets are indexed and their spatial indices can be used to filter out a large portion of

---

[4] http://www.nyc.gov/html/dcp/html/bytes/applbyte.shtml
[5] http://libspatialindex.github.com/
[6] http://www.gdal.org/

candidate pairs to be joined. The filtered pairs are provided as a vector of (fid, tid) pairs where fid is the identifier of a basic unit of the joining dataset (DF) and tid is the identifier of a basic unit of the dataset to be joined (DT). Here the basic unit can be a tree node if tree indexing approaches are used to index the geospatial datasets or a cell of a grid if a grid-file is used for indexing. Gaede and Gunther (1998) and Samet (2005) provide more details on spatial indexing.

In this study we consider joining a point dataset that is indexed by a quadtree and a polygon dataset that is indexed by an R-tree although the proposed approach can be generalized to many other spatial join scenarios. The filter phase can be implemented by querying the R-Tree using the expanded bounding boxes of quadtree nodes, i.e., $(x1 - D, y1 - D, x2 + D, Y2 + D)$ where D is the expansion size. A quadtree node is paired with an R-tree entry if its expanded bounding box intersects with the bounding box of the entry in a leaf R-tree node. Furthermore, we limit our discussion to distance based nearest-neighbor spatial join, i.e., for each point in DF, compute the minimum distance from the point to polygons in DT. Here the distance from a point to a polygon is canonically defined as the minimum distance between the point and all line segments of the rings (including both the outer ring and the inner rings of polygons with holes) of the polygon. While classic quadtree indexing on point data put each point in a tree node, we have observed that for large-scale high-resolution point data, very often points that are close to each other have similar data access patterns and it is beneficial to group them together in spatial joins. For example, there are half a million taxi trip records in NYC per day and there can be thousands of taxi pickup locations roughly at the same point locations. When joining these points and the nearby tax lot polygons (to associate the taxi trips with land use categories that are used as proxies of trip purposes), they will be paired with the same R-Tree entries representing the respective polygons. As such, we assume a leaf quadtree node can hold a set of points. The mechanism can also be viewed as indexing a collection of points (which is supported by OGC Simple Feature Specification as GeometryCollection[7]) using a single quadtree node. The refinement phase requires compute pair-wise distances between the points under a quadtree node and the segments of a polygon indexed by an R-tree. This needs to be done for all the (fid, tid) pairs derived from the filter phase. We further assume that the MBBs of both the quadtree nodes and the R-Tree nodes are provided as vectors and can be accessed by using the fids and tids, respectively. The outputs of the pair-wise distance computation are a vector of triples of (fid + pid, tid + sid, distance), where pid is the identifier of a point in the quadtree node referred by fid, sid is the identifier of a segment of the polygon referred by tid and + denotes concatenation. To find the shortest distance between a point and a polygon, the binary minimum function should be applied to all distances under a same concatenation of fid + pid + tid. We next discuss the layouts of the inputs so that we can concentrate on parallel primitives based spatial join algorithm to be presented in Sect. 5.3.

---

[7] http://www.opengeospatial.org/standards/sfs

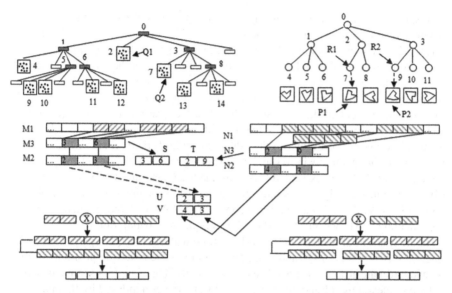

Fig. 5.1 Illustration of spatial join on point (indexed by a quadtree) and polygon (R-tree) and storage layout for pair-wise distance computation

As shown in the middle of Fig. 5.1, each input dataset is organized as three vectors. For the point dataset, the first vector (M1) records the coordinates of points, the second vector (M2) records the numbers of points in the quadtree quadrants and the third vector (M3) records the starting positions of the points in all quadrants correspond to quadtree leaf nodes. Similarly another set of three vectors are used to record the similar information for polygons, i.e., N1 for the segments of polygon rings, N2 for the numbers of polygon segments and N3 for the starting positions of polygon segments in N1. While a vector of (fid, tid) pair can be intuitively used to represent the inputs of the refinement phase, the information stored in M2/M3 and N2/N3 correspond to the fids and tids are copied into four vectors, i.e., U, V, S, T, for fast and convenient data accesses. Here U stores the numbers of points in the quadrants correspond to the fids in the (fid, tid) pairs, V stores the numbers of segments of polygons correspond to the tids in the (fid,tid) pairs, S stores the starting positions of the quadrants in M1 and T stores the starting positions of the polygon segments in N1. As the expanded MBBs of quadtree quadrants many intersect with MBBs of multiple polygons as indicated in the (fid,tid) pairs after the filter phase, both U/S and V/T can contain multiple copies of M2/M3 or N2/N3 elements, but the combinations of (U,V) and (S,T) should be unique in a spatial join.

As shown in the middle of Fig. 5.1, each input dataset is organized as three vectors. For the point dataset, the first vector (M1) records the coordinates of points, the second vector (M2) records the numbers of points in the quadtree quadrants and the third vector (M3) records the starting positions of the points in all quadrants correspond to quadtree leaf nodes. Similarly another set of three vectors are used to record the similar information for polygons, i.e., N1 for the segments of polygon rings, N2 for the numbers of polygon segments and N3 for the starting positions of polygon segments in N1. While a vector of (fid, tid) pair can be intuitively used to represent the inputs of the refinement phase, the information stored in M2/M3 and N2/N3 correspond to the fids and tids are copied into four vectors, i.e., U, V, S, T, for fast and convenient data accesses. Here U stores the numbers of points in the quadrants correspond to the fids in the (fid, tid) pairs, V stores the numbers of segments of polygons correspond to the tids in the (fid,tid) pairs, S stores the starting positions of the quadrants in M1 and T stores the starting positions of the polygon segments in N1. As the expanded MBBs of quadtree quadrants many intersect with MBBs of multiple polygons as indicated in the (fid,tid) pairs after the filter phase, both U/S and V/T can contain multiple copies of M2/M3 or N2/N3 elements, but the combinations of (U,V) and (S,T) should be unique in a spatial join.

In the example shown in Fig. 5.1, Q1 is paired with R1 and Q2 is paired with R2. There are two points in Q1 and three points in Q2. Their starting positions in the point coordinate array (M1) are 3 and 6, respectively. There are four segments in the polygon indexed by R1 and three segments in the polygon indexed by R2 and their starting positions in the polygon segment array (N1) are 2 and 9, respectively. Since the number of pairs after the filter phase is 2, there are two elements in U, S, V and T whose values are [2, 3], [3, 6], [4, 3] and [2, 9], respectively. As shown at the bottom of Fig. 5.1, there will be $2*4=8$ distance computations for the first (fid, tid) pair, $3*3=9$ distance computations for the second (fid, tid) pair, and, there will be $8+9=17$ elements in the resulting distance vector. While it is straightforward to provide a serial implementation on CPUs by using three loops, i.e., the first loop for all (fid, tid) pairs, the second loop for all points in a quadrant and the third loop for all polygon segments, as we shall see in Sect. 5.3, it becomes non-trivial using a parallel computing model as each parallel processing unit (GPU thread) needs to know exactly where to retrieve the data and where to output the results. As such, significant coordination efforts are required which is the core part of the primitives based spatial join algorithm.

Before we present the parallel algorithm in Sect. 5.3, we would like to briefly discuss related works on parallel relational join on both multi-core CPUs (Samet 2005) and GPUs (Blanas et al. 2011; He et al. 2008) which have attracted considerable research interests recently. We note two major differences between relational joins and spatial joins. First, as many spatial indexing approaches (such as R-Trees) allow tree node to have overlapped MBBs, filter based on such spatial indexing structures can result in duplicated tree nodes and subsequently requires non-trivial combinations in the refinement phase. Second, spatial operations are usually more computing intensive than equijoin and most types of theta-joins which make GPGPU accelerations more desirable. Our work is closely related to the seminal work on parallel primitives based spatial joins and geospatial operations nearly 20 years ago (Hoel and Samet 1994). However, while their work targeted for then supercomputers (Thinking Machine[8]) which were only accessible to a few very

---

[8] http://en.wikipedia.org/wiki/Thinking_Machines_Corporation

selective groups, our work targets at the commodity GPUs that are affordable by every researcher now days. In addition, while their work was focused on the filter phase, our research is focused on the refinement phase.

Although it is straightforward to adopt a task-based parallelization schema at the (fid, tid) pair level on CPUs, it might not be a good choice for GPGPUs for two reasons. First, unlike the current generation of CPUs that have a few to a dozen parallel processing cores with each is capable launching 1–2 threads on a single computing core, GPUs typically have hundreds processing cores and support simultaneously launching hundreds of thousands threads. Task-level parallelization many not provide enough parallelism on GPUs for such spatial joins when the number of pairs is less than the number of processing units (blocks or threads) on GPGPUs. Second, the computing loads among the pairs can be highly imbalanced and the task-level parallelization can be too coarse to fully utilize GPGPU computing powers. The skewness can be more severe on GPGPUs than on multi-core CPUs as more processing units on GPGPUs are prone to be underutilized under a same degree of skewness. Furthermore, we are more interested in developing data parallel algorithms that can handle data skewness in an embedded manner than simply passing the burdens of handling skewness to middleware, operating systems or hardware built-in schedule modules which usually do not understand the inherent parallelism among data well.

5.3 Algorithm Design Using Parallel Primitives

While computing efficiency is often the primary motivation for GPGPU applications including spatial joins, we believe it is also important to understand the inherent parallelisms in spatial joins from a research perspective. Parallel primitives that have been implemented in quite a few parallel libraries such as Thrust and CUDPP[9] facilitate quick prototypical implementations. The high-level implementations usually are easy to understand and have better portability although they may not be the most efficient ones that are achievable on parallel hardware. In this study, as an exploratory research effort to understand the parallelisms in spatial joins and their performance on modern GPGPUs, we have decided to adopt a parallel primitive based approach. More specifically, our parallel spatial join algorithm is built on top of the Thrust library that becomes part of Nvidia CUDA SDK after version 4.0. While it is beyond the scope of this paper to introduce the details of the parallel primitives, we refer to the Thrust project website for more technical details on the key primitives, including transform, scan, reduce, gather, scatter and their variants. A brief explanation of the primitives with examples is provided in our technical report.[10]

---

[9] http://code.google.com/p/cudpp/

[10] http://geoteci.engr.ccny.cuny.edu/primquad/primquad_draft.pdf (pages 4–5).

Inputs: Vectors M1, N1, U, V, S, T (as defined in Section 2)
Outputs: Pair-wise distance vector among all points and segments across the K (fid,tid) pairs

1 **Transform** on **U** and **V** to compute the numbers of distance calculation pairs in all (fid,tid) pairs and store the result as vector **X1** using the **multiplies** binary function.
2 **Scan** (**exclusive**) on **X1** to compute the boundary positions of all (fid, tid) pairs and store the result as vector **X2**.
3 **Reduce** on **X1** to compute the total number of distance calculations for all (fid, tid) pairs and store the result as scalar **tot_pairs.**
4 **Scatter** the sequence of (0..K-1) to **X3** using **X2** as the map
5 **Scan** (**exclusive**) on **X3** and store the result in **X4** using the **maximal** binary function
6 **Gather** on **T** using **X4** as the map and store the results to **X5**
7 **Gather** on **S** using **X4** as the map and store the results to **X6**
8 **Scan** (**exclusive**) on **U** and store the result in **X7**
9 **Reduce** on U and compute the number of query points as scalar **tot_points.**
10 **Scatter** the sequence of (0..K-1) to **X8** using **X7** as the map.
11 **Scan** (**exclusive**) on **X8** and store the result in **X9** using the **maximal** binary function
12 **Gather** on **V** using **X9** as the map
13 **Scan** (**exclusive**) on **X10** and store the result in **X11**
14 **Scatter** **X11** using itself as the map and store the result in **X12.**
15 **Scan** (exclusive) on **X12** and store the result in **X13** using the **maximal** binary function
16 **Transform** on a sequence of (0.. **tot_pairs**-1) and **X13** using the **minus** binary function and store the results in **X14.**
17 **Transform** **X5** and **X14** using the **plus** binary function and store the results in **X15**
18 **Scatter** a sequence of (0..**tot_points**-1) to **X16** using **X11** as the map
19 **Scatter** **X7** to **X17** using **X4** as the map.
20 **Transform** on **X16** and **X17** using the **subtract** binary function and store the results in X18
21 **Transform** **X6** and **X18** using the **plus** binary function and store the results in X19
22 **Gather** on **M1** using **X19** as the map and store the results in X20
23 **Gather** on **N1** using **X15** as the map and store the results in X21
24 **Transform** on **X20** and **X21** using a **user-defined point-to-line distance** function and store the results in **X22**
25 **Reduce** (**by key**) on **X22** using the **minimum** binary function and store the results to the **output** vector.

Fig. 5.2 Algorithm design using parallel primitives

The overall structure of the algorithm is provided in Fig. 5.2 and an illustrative example is shown in Fig. 5.3. Although the algorithm looks complex, each of the 25 steps corresponds to a single line of code to invoke a parallel primitive which makes the implementation really simple. The majority of the design (steps 1–21) is devoted to compute the positions of points and segments in their respective storage arrays (M1 and N1) for parallel processing units (CUDA threads in this case) so that distance computation can be performed in parallel (steps 22–25). While it is straightforward to perform a two-level loop in a serial program as the two looping variables

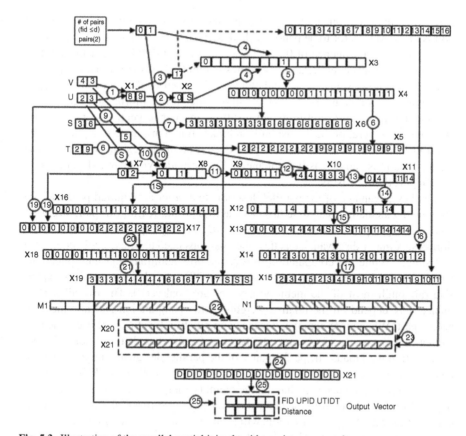

Fig. 5.3 Illustration of the parallel spatial join algorithm using an example

can be increased sequentially, it becomes non-trivial using a parallel computing model. A combination of scatter, scan and gather is required to emulate the sequence of where k loops through all the K (fid, tid) pairs, i loops through all points and j loops through all the segments within a (fid, tid) pair. We next explain how the serial loops can be realized using primitives on parallel machines.

Steps 1–5 builds a template vector to mark the boundaries of (fid, tid) pairs (results stored in X4). Steps 6–7 broadcast the position of the first point in the quadrant corresponds to the fid of the kth (fid, tid) pair to all U[k]*V[k] pairs within the boundary of the pair for all the K pairs (results stored in X5). Steps 8–13 calculate the starting positions of all the points in all the K (fid, tid) pairs. After step 13, vector X11 actually stores $\sum_{k=0}^{P-1}\sum_{i=0}^{U[k]} i * V[k]$ for all possible combinations of k and i. Steps 14–17 compute the positions of segments in the N1 vector to be paired with all points for each (fid, tid) pair. Steps 14–16 essentially generate a sequence of 0 … V[k] − 1 that is duplicated U[k] times within each of the K (fid, pid) pairs to emulate the inner loop. Similarly steps 18–21 compute the positions of points in the M1 vector to be paired with all segments for each (fid, tid) pair. Steps 18–20 generate a sequence of 0 ... U[k] − 1 for each (fid, tid) pair with each element in the sequence duplicated V[k] times to emulate the outer loop. Despite the differences among steps 14–17 and steps 18–21, both of the procedures require broadcast the loop numbers at all levels so that the global information can be used to compute the correct point/segment positions in their respective storage arrays. We note that several of the Thrust built-in binary functions, including plus, minus, multiplies, minimum and maximum, have been used in the design. However, these binary functions can be easily implemented if the underlying parallel libraries do not support them natively.

The key advantage of the parallel design is that, after all required positions are computed, calculating pair-wise distances across all the K (fid, tid) pairs becomes embarrassingly parallelizable and can be easily handled by virtually any parallel libraries. A major disadvantage is that, the positions need to be explicitly computed and stored. While the computation mostly requires additions and multiplications which are very fast on modern GPUs, accessing GPU global memory is very expensive (although most of the memory accesses are coalesced). The overall performance may depend on the relative weights of the computation part and memory access part. When more complex distance computation functions are used, the performance gains can be significant. On the other hand, when there are not sufficient computation workloads, the performance of the parallel design and implementation can be even worse than a straightforward serial implementation on CPUs. A related disadvantage is that the design has a large memory footprint requirement. Although many of the vectors used in the algorithm can be reused in a real implementation, explicitly storing the position vectors will limit the number of (fid, tid) pairs that can be processed on a single GPU device. However, as the design can virtually scale up linearly with the numbers of GPU devices, it is possible to run the implementation in a large cluster computer with fast network connection to address the memory limitation problem.

5.4 Implementation and Experiments

While our initial research is driven more by algorithmic design, to gain more insights on its performance for practical applications, we have implemented the algorithm on top of the Thrust library. The source code is available in our website.[11] As discussed previously, the implementation is fairly straightforward using Thrust where each step is implemented as a call to the respective Thrust primitive. The implementation is compiled using CUDA SDK 4.0 and the experiments are performed on an Nvidia Quadro 6000 GPU with 448 cores (1.15 GHz) and 6 GB memory.[12]

---

[11] http://geoteci.engr.ccny.cuny.edu/primspjion/primspjoin.htm

[12] http://www.nvidia.com/object/product-quadro-6000-us.html

The host machine is a Dell Precision T5400 with dual Intel E5405 CPUs (2.0 GHz) and 16 GB memory. For point data, we use a 1 % sample of the pickup locations of taxi trips in Manhattan in January 2009 as the point data which has 122,043 point locations. For polygon data, the MapPluto Tax Lot data in Manhattan is used which has 43,252 polygons and 293,335 vertices. Many of the tax lot polygons have regular shapes. This explains that the average number of vertices per polygon is only around 7. However, a small portion of the polygons have hundreds of vertices which make the polygon data skewed in computation. We have empirically set the maximum number of points in a quadtree quadrant (P) to 100 and set D to 100 ft for expanding bounding boxes in the filter phase.

As a comparison, we have also implemented the refinement phase on CPUs using straightforward loops. Our data storage layout design discussed in Sect. 5.2 actually fits CPU hardware architectures very well. Looping through one dimensional arrays is naturally cache friendly. Given that accesses to memory can be hundreds of times slower than accesses to registers and CPU architectures increasingly rely on caching to improve memory access performance, it is important to be cache friendly. Unlike the parallel primitives based GPU implementation that requires computing, writing and reading positions and distances explicitly to GPU memory in order to use the parallel primitives, they can reside in CPU registers which is more efficient. For fair comparison, we have put both the CPU and primitives based GPU implementations in a same program and used –O3 flag to optimize for speed for the CPU code. We note that both implementations use a same filter phase and thus we will only compare the performance of the refinement phase. Our results showed that the end-to-end runtime of the primitives based GPU implementation is 345.75 ms to join the 122,043 points in 8,447 quadrants and 43,252 polygons. A total of 142,927,001 distance computation for 1,503,640 (fid, tid) pairs has been observed. In contrast, the optimized CPU implementation requires 2,325.84 ms. As such, a 6.7× speedup is achieved for the primitives based GPU implementation.

Both the CPU and GPU based implementation are main-memory based using the storage layout discussed in Sect. 5.2. As existing spatial databases and GIS are mostly designed for disk-resident data, to understand how main-memory based systems can improve the performance of geospatial computing, we have implemented the same spatial join task using open source packages. The open source based implementation uses libspaitalindex for R-tree indexing and query processing in the filter phase and GDAL/OGR for distance computation between points and polygons. For each point with a coordinate of (x,y), we first query the R-Tree to extract polygons that are with a square of $(x-D, y-D, x+D, y+D)$ and then compute the minimum distance from the point to the polygons. Our experiments have shown that the end-to-end runtime of the open source implementation is 88,531.18 ms which is 38 times slower than the CPU implementation and 256 times slower than the primitives based GPU implementation. The results clearly demonstrate the efficiency of main-memory based implementations. We believe the performance gap between disk-resident and main-memory based implementations can be attributed to the following factors. First, traditional Spatial Databases and GIS assume a very limited

CPU memory capacity and uses sophisticated data structures and algorithms to accommodate memory limitations. However, given the increasing memory capacities (tens of GBs to TBs) and decreasing prices (~$5/GB retail prices), the assumed limitation is not valid anymore. Second, pointers and dynamic memory allocations have been extensively used in many spatial database and GIS software developments and they may not be cache friendly which is becoming increasingly important in modern hardware architectures including both CPUs and GPUs.

Based on the experiments and analysis, we suggest re-examining the performance bottlenecks in geospatial computing by adopting an integrated hardware and software co-design approach. As computer processors are evolving into a parallel era, it is essential to fully utilize the parallel computing power provided by multi-core CPUs, many-core GPUs and distributed computing nodes (Zhang 2010). Existing Spatial Databases and GIS software need to be adapted to the new hardware architectures to efficiently process large-scale geospatial data and effectively solve real world problems. As a first step towards the adaptation, understanding the inherent parallelisms in major geospatial computing algorithms and designing their parallel implementations on top of well-understood and well-supported parallel primitives can be important. They can lay solid foundation in developing spatial-specific parallel primitives that are both high-performance and easy to use for geospatial computing applications.

5.5   Conclusion, Discussion and Future Work

We have designed a parallel spatial join algorithm that is suitable to implement on parallel machines including GPGPUs. Our prototypical implementation using the Thrust parallel library has demonstrated a 6.7× speedup over an optimized CPU serial implementation. The result is encouraging in the sense that native implementations of spatial joins directly on top of GPU accelerators can potentially achieve much higher speedups for spatial joins which are fundamental to Spatial Databases and vector GIS.

From a methodological perspective, the serial CPU implementation and the parallel primitives based GPU implementation represent two extremes with respect to efficiency and scalability. The serial implementation is efficient but not scalable for parallel execution while the primitives based implementation is scalable but not very efficient due to memory access overheads in storing and retrieving positions and intermediate results. We believe some hybrid approaches can potentially achieve both high efficiency and scalability at the levels that are appropriate for applications. For example, for polygon data that are not extremely skewed, it might be more beneficial to use two-levels of parallelisms on CUDA-enabled GPUs, i.e., (fid, tid) pairs at the computing block level and pair-wise distance computation at the thread level. In this case, both points and polygon segments correspond to a (fid, tid) pair can be loaded to GPU shared memory by all the threads in a computing block collaboratively so that no direct global memory accesses are needed during distance computation. Computing within a GPU computing block can be much similar to the CPU serial implementation with respect to the two-level loop for the pair-wise distance computation within a (fid, tid) pair.

For future work, first, we would like to first implement the hybrid idea to explore design tradeoffs and potential performance gains. Second, we plan to implement a few indexing algorithms on GPGPUs for the filter phase so that we can perform spatial joins completely on GPUs. Third, while the framework of the spatial join algorithm discussed in this paper is generic, our current implementation is limited to joining points with polygons. We plan to make the implementation more generic to accommodate spatial joins of other data types for both parallel primitives based and hybrid designs.

Acknowledgment This research was supported partially by the PSC-CUNY grant 65692-00 43.

References

Blanas, S., Li, Y. and Patel, J., 2011. Design and evaluation of main memory hash join algorithms for multi-core CPUs. Proceedings of ACM SIGMOD Conference.

Clematis, A., Mineter, M. et al., 2003. High performance computing with geographical data. Parallel Computing 29(10): 1275–1279.

Gaede V. and Gunther O., 1998. Multidimensional access methods. ACM Computing Surveys 30(2): 170–231.

He, B., Yang, K., et al. 2008. Relational joins on graphics processors. Proceedings of ACM SIGMOD conference.

Hoel, E. G. and Samet, H., 1994. Performance of Data-Parallel Spatial Operations. Proceedings of VLDB Conference, 156–167.

Hong, S., Kim, S. K., et al., 2011. Accelerating CUDA graph algorithms at maximum warp. Proceedings of the 16th ACM symposium on Principles and practice of parallel programming, 267–276.

Hou, Q., Sun, X., et al., 2011. Memory-Scalable GPU Spatial Hierarchy Construction. IEEE Transactions on Visualization and Computer Graphics 17(4): 466–474.

Jacox, E. H. and Samet, H., 2007. Spatial join techniques. ACM Transaction on Database System 32(1), Article #7.

Luo, L., Wong, M. D. F., et al., 2011. Parallel implementation of R-trees on the GPU. Proceedings of the 17th Asia and South Pacific Design Automation Conference (ASP-DAC), 353–358.

Samet, H., 2005. Foundations of Multidimensional and Metric Data Structures Morgan Kaufmann.

Zhou, K., Gong, M., et al., 2011. Data-Parallel Octrees for Surface Reconstruction. IEEE Transactions on Visualization and Computer Graphics 17(5): 669–681.

Zhang, J., You, S. and Gruenwald, L., 2010. Indexing large-scale raster geospatial data using massively parallel GPGPU computing. Proceedings of ACM-GIS, 450–453.

Zhou, K., Hou, Q., et al., 2008. Real-Time KD-Tree Construction on Graphics Hardware. ACM Trans. on Graphics 27(5).

Zhang, J., 2010. Towards Personal High-Performance Geospatial Computing (HPC-G): Perspectives and a Case Study. Proceedings of ACM HPDGIS workshop, 3–10.

Chapter 6
Utilizing CUDA-Enabled GPUs to Support 5D Scientific Geovisualization: A Case Study of Simulating Dust Storm Events

Jing Li, Yunfeng Jiang, Chaowei Yang, and Qunying Huang

Abstract Visualizing geospatial data from model simulations and field observations is an important way to support scientific data explorations. Compared to general scientific visualization, visualizing 5 Dimensional (5D) geospatial data possesses computational challenges of handling large volume unstructured 5D datasets. Utilizing state-of-the-art computing techniques is essential in developing an efficient visualization pipeline. The recent emerging Compute Unified Device Architecture (CUDA)-enabled Graphics Processing Units (GPUs) can provide a potential solution to address computational intensity by offering parallel computing capabilities. This chapter describes our efforts of designing and implementing a CUDA-based visualization pipeline for 5D geospatial data stored in scientific data formats. Such pipeline includes four major steps: data filtering, coordinate transformation, interpolation and rendering. Except for data filtering, all other three steps have been implemented with CUDA-enabled GPUs. Based on the implementations and performance tests, we summarize the major advantages and disadvantages of employing GPUs to implement 5D scientific geovisualization.

Keywords Geovisualization • Geospatial Cyberinfrastructure • CyberGIS • EarthCube • Big Data

J. Li
Department of Geography and the Environment, University of Denver, Denver, CO, USA
e-mail: Jing.Li145@du.edu

Y. Jiang
Department of Geography and Geoinformation Science, Joint Center for Intelligent Spatial Computing, George Mason University, Fairfax, VA, USA
e-mail: yjiang7@gmu.edu

C. Yang (✉) • Q. Huang
Department of Geography and Geoinformation Science,
George Mason University, Fairfax, VA, USA
e-mail: cyang3@gmu.edu; qhuang1@gmu.edu

X. Shi et al. (eds.), *Modern Accelerator Technologies for Geographic Information Science*, DOI 10.1007/978-1-4614-8745-6_6, © Springer Science+Business Media New York 2013

6.1   Visualization in Geospatial Sciences

Geovisualization is a research area that deals with the concepts, methods and tools of visualizing geospatial data (MacEachren and Kraak 2001). Through applying graphical representation techniques on data, scientists can understand the structures of geographic features, identify the driving forces of dynamic phenomena and thus determine the relations between multiple spatial variables. As generic visualization methods and tools do not take the special characteristics of geospatial data into consideration, scientists have been customizing these methods and tools to support scientific explorations of such data. For example, to visualize datasets in different coordinate systems in the same virtual environment, scientists have developed geometric transformations to facilitate the transformation of geospatial features between different coordinate systems.

If we classify geovisualization based on nature of objects that are visualized, two general forms can be identified: information visualization and scientific visualization (Card et al. 1999). Information visualization represents the abstract and non-physical characteristics and properties of data, for example, the relationships between multiple attributes. In contrast to information visualization in geospatial sciences, scientific visualization focuses on the representation of scientific data that capture the geophysical processes such as ocean currents, movements of hurricanes and fluid dynamics. Scientific visualization usually gives the initial visual representations of data which help formalize hypothesis to be tested through information visualization.

Over the past few decades, various methods, techniques and tools have been designed and developed in supporting visualization of geospatial data. Notable open source visualization tools include VisAD (Hibbard et al. 1996), World Wind (Bell et al. 2007), GeoViz Toolkit (Hardisty and Robinson 2011) and Integrated Data Viewer (IDV, McKaskle and Rogers 2009). Through providing multiple visualization functions, these tools have greatly enhanced our capabilities in exploring the hidden patterns captured by complex geospatial data of different spatiotemporal scales. However, providing an efficient visualization solution for representing large scale massive 5D geospatial data is still challenging.

6.2   5D Scientific Geovisualization

Geophysical processes are described by 5D data from model simulations or observations. By "5D", we mean three spatial dimensions, one temporal dimension and one dimension for thematic attribute. As such data are featured with large volume, multidimensional and high degree of complexity in space and time, interactive scientific visualization has been employed as a mean to manipulate the data and view the processes visually (e.g., *Vis5D*, Hibbard et al. 1996). To visualize 5D data capturing dynamic volumetric phenomena, a typical strategy is to implement volumetric rendering algorithms on the 4D data extracted at different snapshots to generate a series of visual products (e.g., images for ray casting). The coloring schema is determined by the range of the chosen attribute to be displayed. With the visual products sorted by their temporal stamps, animation is triggered to play the visualization in sequence.

Three types of bottlenecks can significantly impact the overall performance of the visualization pipeline (Yang et al. 2011). First, due to the large volume of 5D datasets, which can easily exceed the maximum memory of computing facilities, it is infeasible to load all data at one time. Out-of-core solutions should be designed to improve the I/O efficiency and therefore handle large volume datasets. Second, processing data and implementing geometric calculations can be computationally intensive. The computational intensity is largely determined by the data volume and the chosen algorithms for visualization and the intensity can be greatly increased when large volume data are involved. High performance computing techniques should be incorporated to address the computing intensity. Third, real time interactive rendering requires the implementation of rendering algorithms and displaying the visual result within a predefined time interval (e.g., 0.3 s for frame, Guthe et al. 2002). Graphics hardware accelerations are necessary to enhance the interactivity.

6.3 Potentials of Using GPUs to Support Scientific Geovisualization

To address the computing intensity in scientific geovisualization, scientists have implemented parallel computing models with multi-core Central Processing Units (CPUs) (e.g., Eilemann et al. 2009). Recently, the emerging Compute Unified Device Architecture (CUDA)-enabled Graphics Processing Units (GPUs) have greatly altered the parallel computing paradigm, which can contribute to the development of an efficient scientific geovisualization platform for 5D data. The CUDA is "a parallel computing platform and programming model" that allows users to manage the computing power of GPUs through implementing its C style programmable interface (NVIDIA 2007).

CUDA has the following characteristics: (a) it shows better floating point (FP) performance when applied both general arithmetic calculations and rendering scenarios; (b) its threading model allows users to distribute functions, which are called kernels in GPU, to the multiple blocks and threads through two-level multi-threading technique; (c) through offering interfaces to various programming languages such as C++ and Java, it ensures parallel execution on multiple GPU cores and seamless integration with existing platforms; (d) compared to traditional GPU programming interfaces which are designed for graphical applications (e.g., C for graphics, Cg), CUDA supports a broad range of general purpose GPU (GPGPU) functions. The modern GPU-enabled graphics card is not only a powerful graphic rendering device but also a highly parallel programmable processor featuring peek arithmetic and memory bandwidth that substantially counterpart CPU. Therefore, in this research, we explored how CUDA-enabled GPUs can change the visualization pipeline and improve the overall efficiency.

6.4 Implementations of a GPU-Based Geovisualization Framework

In designing the 5D geovisualization framework, we chose NASA World Wind as the virtual geographic environment (Fig. 6.1). World Wind serves as a virtual geographic environment in the visualization framework, which provides the base remote sensing images and terrain models. As an open source visualization platform, World Wind provides interfaces for users to customize rendering functions for other types of geospatial data, for example, 5D geospatial data. In this way, users can better understand the interaction between the dynamic phenomena and existing geographic settings.

The 5D geovisualization pipeline consists of the following major modules: data filtering, coordinate transformation, interpolation and sampling, and volumetric

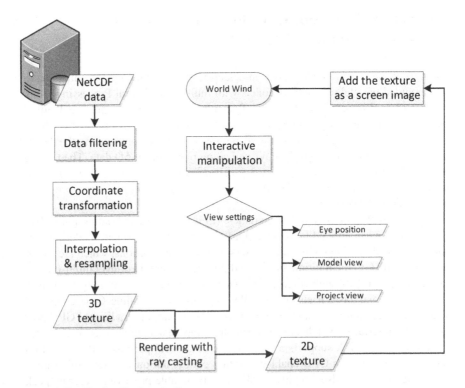

Fig. 6.1 The 5D visualization pipeline

rendering and display. Data filtering facilitates fast access geospatial data to be delivered to the visualization pipeline. This involves a spatiotemporal query that filters the data stored in scientific data formats (e.g., NetCDF, HDF-EOS) within a spatiotemporal coverage and loads filtered data into the main memory. Coordinate transformation and 3D interpolation are then performed on the filtered data to generate a series of regular 3D textures as required by the ray casting process. The filtered 5D data array at a timestamp corresponds to a 3D texture. If users are interested in a specific timestamp, ray casting algorithm is performed on the 3D texture corresponding to that timestamp only. Once the ray casting process finishes, a 2D texture generated from the 3D texture is directly attached to the World Wind as a screen image layer. To view a dynamic volumetric process, the animation mode is enabled. Time series 3D textures are sent to the ray casting module and the ray casting module performs calculations on 3D textures in a sequential manner. As World Wind provides interfaces that allow users to manipulate the geographic features interactively, the ray casting will be executed whenever the view settings are updated. As a result, the computational and the rendering intensity can increase significantly.

We implemented coordinate transformation, interpolation and resampling, and ray casting with CUDA. Although data filtering can be memory intensive, we implemented the process with host programs on CPUs because CUDA does not provide native interfaces to support reading/writing scientific data formats. However, the computational intensity of the rest three intensive tasks has been shifted to GPUs. Below, we will describe our implementations of visualization pipeline with a focus on three major processes which are coordinate transformation, interpolation and ray casting.

Before discussing the implementation in detail, we first briefly review the parallel computing framework of CUDA, which is the foundation of our parallel model design. CUDA offers a two-level parallel hierarchy: thread-level and block-level. Threads are grouped into blocks and blocks are grouped into a grid. The number of threads within a block can be specified with 1D (x), 2D (x, y) or 3 D (x, y, z). Similarly, the number of block is specified with 1D (x) and 2D (x, y). The total number of threads is equal to the value of the number of threads within a block multiplies the total number of blocks. A kernel, similar to a function of CPU programming, is executed by an array of threads of a grid. For example, a kernel can be a coordinate transformation performed on every element of a data array. With the data parallelism, given the configurations of the block size of a grid and the thread size of a block, the data array can be equally divided into sub-regions with smaller dimensions and these sub-regions are distributed to different threads. Then each sub-region is processed by a number of threads within a block. Each thread has a unique ID that it uses to access the sub-region to the thread. In this way, the total execution time can be reduced through the parallel processing.

In the context of 5D geovisualization, the time varying volumetric data can be treated as time series 3D arrays. We set up 3D thread-blocks and divided the 3D arrays along three spatial dimensions. For each cell from the data array, the program derives its geographic coordinates and transfers geographic coordinates into projected coordinates according to the view settings (Fig. 6.2). The transformation

```
///cuda.cu
#include <cutil_inline.h>
///@kernel function, Transform, which is used for transforming spatial information between different coordinates system.
__global__ void Transform(double* A, double* B, double* C, int* BOUNDARY)
{
        ///get data index in array using thread id, block id and block dim.
              int i = blockDim.x * blockIdx.x + threadIdx.x;
        int j = blockDim.y * blockIdx.y + threadIdx.y;
        int k = blockDim.z * blockIdx.z + threadIdx.z;
        int idx = i * BOUNDARY[1] * BOUNDARY[2] + j * BOUNDARY[2] + k;

        /// transform data from Geo-coordinate system (latitude, longitude and    ///elevation) into Cartesian coordinates system
        double equatorialRadius = WGS84_EQUATORIAL_RADIUS;
        double es = WGS84_ES;
        double cosLat = cos(A[idx] / 57.3);
        double sinLat = sin(A[idx] / 57.3);
        double cosLon = cos(B[idx] / 57.3);
        double sinLon = sin(B[idx] / 57.3);

        double rpm = equatorialRadius / sqrt(1.0 - es * sinLat * sinLat);
        double x = (rpm + C[idx]) * cosLat * sinLon;
        double y = (rpm * (1.0 - es) + C[idx]) * sinLat;
        double z = (rpm + C[idx]) * cosLat * cosLon;

        A[idx] = x;
        B[idx] = y;
        C[idx] = z;
}
```

Fig. 6.2 The kernel function of coordinate transformation

```
__global__ void Interpolate(int* DATA)
{
        ///get data index in array using thread id, block id and block dim.
        ///with three dimensions.
        int i = blockDim.x * blockIdx.x + threadIdx.x;
        int j = blockDim.y * blockIdx.y + threadIdx.y;
        int k = blockDim.z * blockIdx.z + threadIdx.z;

        int index = i * d_volumeSize * d_volumeSize + j * d_volumeSize + k;
        ///calculate current data value by nine neighboring points.
        for(int ii = i-1; ii < i + 2; ii++)
        {
          for(int jj = j-1; jj < j + 2; jj++)
          {
                for(int kk = k-1; kk < k + 2; kk++)
                {
                      if( ii > 0 && ii < d_volumeSize && jj > 0 && jj < d_volumeSize && kk > 0 && kk < d_volumeSize)
                      {
                             DATA[ii * d_volumeSize * d_volumeSize + jj * d_volumeSize + kk] = DATA[index];
                      }
                }
          }
        }
}
```

Fig. 6.3 The kernel function of interpolation

equation is provided by World Wind Java Software Development Kit (SDK). Three CUDA arrays, A, B and C are initiated to store x, y and z coordinates respectively. This is an implementation of data parallelism.

We used a simple neighborhood interpolation algorithm to generate 3D textures. Given the projected coordinates of cells of the 3D array, the interpolation first maps each cell to the corresponding voxel within the 3D texture and then fills the empty voxels with the values from neighborhood cells. We applied the same parallel strategy as the coordinate transformation to assign the value of a cell (Fig. 6.3).

Compared to the first two implementations, ray casting is a more complicated process that interactively checks intersections between rays and a 3D texture. Here we only describe the implementation of the core function, "d_render", which

```
__global__ void
d_render(uint *d_output, uint imageW, uint imageH,
         float density, float brightness,
         float transferOffset, float transferScale)
{
    const int maxSteps = 100;
    const float opacityThreshold = 0.95f;

    uint x = blockIdx.x*blockDim.x + threadIdx.x;
    uint y = blockIdx.y*blockDim.y + threadIdx.y;
    if ((x >= imageW) || (y >= imageH)) return;
    // calculate eye ray in world space
    Ray eyeRay;
    //set up view port
      eyeRay.o.x = eye_ori[0];
      eyeRay.o.y = eye_ori[1];
      eyeRay.o.z = eye_ori[2];

    //calcuate the dir
    double a[3],b[3];
      unproject(x, y, 0, c_ViewMatrix, c_ProjMatrix, imageW, imageH, a );
      unproject(x, y, 1, c_ViewMatrix, c_ProjMatrix, imageW, imageH, b);
    eyeRay.d = normal3(subtract(b,a));

    //find intersection with box
    double tnear, tfar;
    int hit = intersectBox(eyeRay, &tnear, &tfar);
    if (!hit) return;
    if (tnear < 0.0) tnear = 0.0;       // clamp to near plane
    double tstep = (tfar-tnear)/maxSteps;
    // march along ray from front to back, accumulating color
    float4 sum = make_float4(0.0f);
    float t = tnear;
    double3 posNow = add(eyeRay.o, mul(eyeRay.d,tnear));
    double3 step = mul(eyeRay.d,tstep);
    for(int i=0; i<maxSteps; i++) {
        // read from 3D texture
        // remap position to [0, 1] coordinates
    double3 pos = add(posNow, mul(step,i));
        float sample = tex3D(tex, (float)((pos.x-boxMinArray[0])/(boxMaxArray[0]-boxMinArray[0])), (float)((pos.y-
boxMinArray[1])/(boxMaxArray[1]-boxMinArray[1])), (float)((pos.z-boxMinArray[2])/(boxMaxArray[2]-boxMinArray[2]))));

// calculate colors
.........skipped here........
{color functions}

        // exit early if opaque
        if (sum.w > opacityThreshold)
            break;

        t += tstep;
        // if (t < tfar) break;


    }
    sum *= brightness;
    // write output color
    d_output[y*imageW + x] = rgbaFloatToInt(sum);
}
```

Fig. 6.4 The kernel function of "d_render"

performs the ray-object intersection for the entire 3D texture (Fig. 6.4). As the source codes provided by CUDA SDK are designed to implement ray casting within a unit cube, we modified the ray intersection portion so that the algorithm can be implemented on any regular 3D texture stored in coordinates in a predefined visualization environment.

To demonstrate the visual effects, we performed the visualization on regional dust storm data generated from the DREAM Eta 8-bin model. The data are stored in NetCDF format and the dust density variable was extracted to characterize the dust storm (Table 6.1). The view frame is 783 pixels in width and 584 pixels in height. Given the original spatial dimensions of the data, a 3D texture with a size of 512^3 is

Table 6.1 Metadata of the sample dust density data

Parameter	Range	Dimension
Longitude	25.99°–36.01°	242
Latitude	−107.97°–113.01°	481
Elevation	0–5,000 m	24

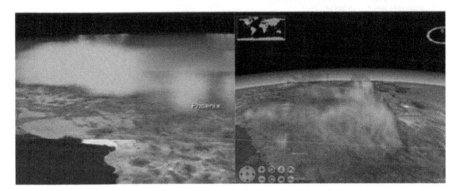

Fig. 6.5 Visual effects of the volumetric rendering from two different view angles

defined to represent the 3D data. Although the texture size is considered as moderate, this is the maximum texture size given the memory of GPUs used in the experiment. Figure 6.5 shows the visual representation when the data were rendered in the World Wind.

6.5 Challenges of Adopting GPUs in 5D Scientific Geovisualization

While CUDA provides relatively high level programming interfaces, the implementation of the visualization pipeline is not straightforward. Three major challenges are (a) integrating kernels with World Wind Java SDK; (b) fine tuning the parallel computing configuration to achieve best performance and (c) designing and implementing data management and communications between CPUs and GPUs in handling multidimensional geospatial data.

6.5.1 Integrations with Existing Visualization Platform

In this framework, the World Wind Java SDK was selected as the visualization development platform. Typical CUDA kernels written in C codes cannot be directly executed on this platform. Therefore, additional configurations and developments are required to address the compatibility issue. In this research, two approaches were designed to kernels with the World Wind Java SDK, which are Java Native Interface (JNI) and the Java bindings for CUDA (JCUDA). JNI is a traditional method used to incorporate C/C++ functions into a Java program. With JNI, CUDA kernels written by in C language can be directly invoked or called by the World Wind visualization application. As a third party library, JCUDA is a complete implementation of CUDA interfaces with Java and thus is frequently used as an alternative solution to call CUDA functions within a Java environment. Such binding allows users to interact with CUDA interfaces (e.g. CUDA Runtime API) while kernels should be precompiled to be used in JCUDA.

Two approaches offer different levels of customizations and interactivities. Given the abundance of Java-based scientific geovisualization tools and software (e.g. VisAD), developing a set of high level cross-platform interfaces built upon the functionalities provided by CUDA is essential to improve the performance of these tools in handling computing intensive tasks. Using the two approaches described above, all interfaces provided by CUDA can be redesigned and packed in a Java development environment as a library. This extends CUDA capabilities by offering cross-platform wrappers. As a result, issues caused by incompatibilities among development platforms and programming languages can be handled by such wrapper.

6.5.2 Parallel Computation Setup and Configurations

The configurations of grid size and block size can change the performance significantly. The specifications of a CUDA-enabled GPU graphics card usually provide information regarding the maximum threads, for example, the maximum active threads. When tuning the parallel configurations, these parameters should be considered. In addition, in executing the kernels, the block size should be a multiple of 32 threads (1 warp) because CUDA triggers the kernels following a *SIMT* (Single Instruction, Multiple Thread) fashion (Nickolls et al. 2008). Taking ray casting rendering algorithm as an example, if the resolution of screen is $1,024 \times 1,024$, performing ray casting rendering with a block size as 17×17 is usually faster than that with a block size as 32×32.

However, such condition may not be always satisfied because the grid size and the block size are constrained by the dimensions of gridded geospatial data. In Table 6.1, any dimension of the data array is not a multiple of 32. Therefore, a few blocks are assigned less data cells. Such inequality leads to the imbalance of tasks between different threads and further increases the total execution time. To explore the appropriate configurations of block size and grid size, we conducted an experiment with a desktop equipped one CUDA-enabled GPU graphics card. Table 6.2 summarizes its hardware configurations and benchmark results. As the selected GPU has two Streaming Multiprocessors (SM), we also enabled two physical cores of the CPU to compare the performance of both computing facilities. The volume of

Table 6.2 Benchmark testing results of the selected computing equipment

	CPU	GPU
Brand	Inter	NVIDIA
Model	i7, 920	GT 430
Cores per unit	2	2 Multiprocessors
Clock	2,660.01 MHz	1,400 MHz
Memory	4,096 M + 2,047 MB	1,023.668 MB
Global memory read	10.685GB/s	13.750GB/s
Global memory write	7.919 GB/s	19.765 GB/s
Global memory copy	9.374 GB/s	19.203 GB/s
Local memory bandwidth	149.252 GB/s	Float-Aligned 171.815GB/s
Float ops(add)	16.917 GFLOPS	127.592 GFLOPS
Double ops(add)	8.435 GFLOPS	11.014 GFLOPS
Image processing brightness histogram	790.005 MegaPixels/s	765.799 MegaPixels/s

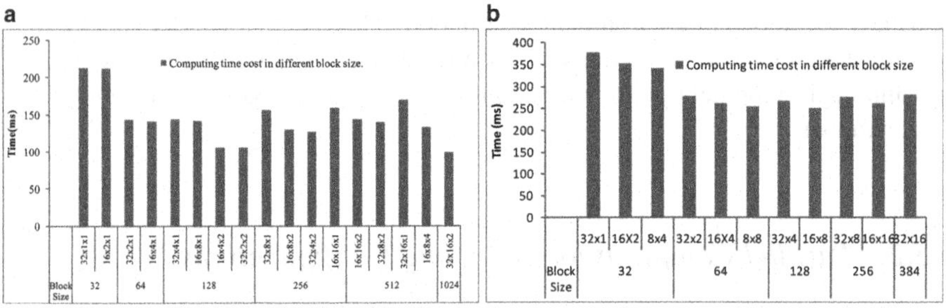

Fig. 6.6 Time costs of executing kernel functions

main memory of CPU does exceed that of device memory of GPU and the clock speed of the CPU is also higher than that of GPU. However, most GPU-based operations (both memory and mathematical calculations) except for the image processing have better performance compared to CPU-based operations.

We examined the impacts of parallel configurations in influencing the computing performance through tuning the block size defined in the kernels of coordinate transformation and ray casting (Fig. 6.6). Two kernels employ 3D blocks and 2D blocks respectively. The interpolation kernel is not evaluated because it has a similar parallel configuration as coordinate transformation kernel.

Figure 6.6 shows that the time cost decreases with the number of threads within each block in general. Therefore, we can usually assign a large number of threads within each block size so that all computing units of the GPU can be activated to reach the peak of computing performance. The minimal value of time cost appears at the point with 1,024 threads which is the maximal number of threads of a block given the configuration of the selected GPU. However, we do not always obtain performance gains with a larger number of threads.

In addition, given the same number of threads allocated to a block, the performance varies with the configurations of a block. In the case of coordinate transformation, every thread executes the transformation for one cell from the data array. If we evaluate the performance with respect to the dimensions of a block and the dimension of the original data, the dimensions of a block determine the total number of blocks of the grid used to execute coordinate transformation. For example, 11,712 blocks are needed given the block size $32 \times 8 \times 1$ whereas 11,616 are needed given the block size $32 \times 4 \times 2$. A few blocks will be assigned cells less than the number of threads within a block. As a result, time costs of two configurations change slightly. In the case of ray casting, the situation is more complicated. While the configuration of a block leads to the variations of total number of blocks (7,300 blocks for 32×2 vs. 7,154 blocks for 8×8), the life cycles of all threads are not identical. The ray-data intersection can be terminated depending on the values of cells along that ray. Consequently, the runtime of a thread executing the ray-data intersection can significantly differ from another thread within the same block. It is hard to predict the total execution time. Therefore, further explorations should be done to identify an appropriate configuration of a block to obtain the best performance.

6.5.3   Data Management and Communications Between GPUs and GPUs

The communication between CPUs and GPUs plays an important role in the CUDA programming model as the latency caused by communication directly affects the performance of computing. In designing the visualization pipeline, we found inter device communications between host (CPU) and device (GPU) may occur at three different stages: (a) as the CUDA API does not provide native support for the scientific data formats (e.g., NetCDF) and the kernels can be executed on the GPU, data stored in these formats should be preprocessed with CPU functions and sent to GPU kernels. If the spatiotemporal coverage used for data filtering does not change, such preprocessing is one time only and therefore the data transfer is one time only operation; (b) the ray casting is triggered whenever view settings of the visualization environment are updated. Interactive manipulations on the data change the view settings continuously. Any changes of the view settings of World Wind should be notified at GPU side immediately to invoke the ray casting process; (c) 2D textures generated from the ray casting process can be sent back to CPU to generate 2D images. The first two communications are mandatory whereas the last communication is optional.

To design CPU-GPU data communication components in the pipeline, we should refer to the memory hierarchy of CUDA because the memory hierarchy determines the types of variables used to facilitate communications. CUDA threads can access data from multiple memory spaces during their executions. Each thread can access private local memory. Shared memory is accessible to all threads of the same block whereas global memory is shared by all blocks within a grid (CUDA 2012). As the CUDA programming model assumes that the CUDA threads execute kernels on a physically separate device and both host and device maintain own separate memory spaces, a program should be developed to manage all memory spaces and facilitate data transfers between CPU and GPU. Data residing on CPU should be copied to GPU to be used for executing kernels with a CUDA memory copy function.

Given different scopes of variables, global variables should be used to store scientific data and view settings (i.e., projection matrix, model matrix and eye positions). Through copying the information from host to device, all threads can access the information to perform ray casting. Except for scientific data, variables of view settings should be consistently updated whenever interactive manipulation occurs. Similar to the ray casting example provided by NVIDIA SDK, our ray casting host program allocates device memory to store the 2D texture generated from ray casting. The texture is directly mapped to a Pixel Buffer Object (PBO) as final display. While global variables are less efficient compared to other variables such as shared variables, these variables are probably the primary variables used to facilitate the data communication between CPU and GPU.

6.6   Conclusions and Future Work

GPU-based computing and rendering techniques have been adopted in multiple application domains. In this study, we explored the usages of CUDA-enabled GPUs to support scientific visualization for multidimensional geospatial data. We found that due to the different designs of parallel frameworks, GPU-based implementations and CPU-based implementations and corresponding taxonomies are significantly different. For example, functions are called kernels with the CUDA. It is highly recommended that scientists should comprehend the programming architecture before designing CUDA programs. While our implementations show the potential of using CUDA to improve the efficiency of 5D scientific geovisualization, current research focuses on implementing the visualization pipeline with the computing resources provided by one desktop machine. To optimize the pipeline, future work includes the following aspects (Li et al. 2013).

First, an optimal algorithm should be designed for parallel configuration. According to the framework, CUDA usually requires two-level parallelism: block-level and thread-level. Users have to tune the configurations of blocks and threads within a block to obtain the best performance, meanwhile, avoid the invalid parallelism settings (e.g., the total number of threads exceeds the total number of voxels). While we can always obtain performance gains with GPU-based implementations by invoking the data and/or task parallelisms, the performance gains can vary with the types of variables used to execute programs and the parallelism configurations. In our experiment, we found that different thread settings can alter the performance of executing parallel programs. The optimal algorithm should automatically identify the best parallel configuration given the data and kernels.

Second, a concrete data management and communication plan should be proposed. The communication between CPUs and GPUs plays an important role in the visualization pipeline. We used global variables to facilitate the communication. While global variables should be less frequently used because of the low performance, such type of variables is essential in the communications between CPUs and GPUs (e.g., data and message transfer). However, shared variables and device functions can be introduced to kernels to promote the communication within a data block. Future research should include replacing unnecessary global variables with shared variables.

Third, distributed computing paradigm should be introduced to enhance the capabilities of the current visualization pipeline in handling "Big Data". Due to the limitation of GPU memory, we only tested the medium sized data. The tendency is that large volume data, advanced computing resources and multiple display devices are distributed within a network environment. Investigating GPU clusters or clouds, which built upon scalable high performance platforms, is critical in enhancing the capabilities of the visualization pipeline to visualize massive 5D geospatial data. In such highly distributed environment, web-based rendering techniques such as WebGL should be incorporated to leverage the rendering and computing capabilities between servers and clients.

Acknowledgements Research and development reported is partially supported by NSF (IIP-1160979 and CNS-1117300) and NASA (NNX12AO34G).

References

Bell DG, Kuehnel F, Maxwell C, Kim R, Kasraie K, Gaskins T, Hogan P, Coughlan J (2007) NASA World Wind : Opensource GIS for Mission Operations, In: Proceedings of IEEE Aerospace Conference:1–9

Card SK, Mackinlay JD, Shneiderman B (1999) Readings in Information Visualization: Using Vision to Think Morgan Kaufmann, San Francisco, CA

Eilemann S, Makhinya M, Pajarola R (2009) Equalizer: A Scalable Parallel Rendering Framework. IEEE Transactions on Visualization and Computer Graphics 15:436–452

Guthe S, Wand M, Gonser J, Strasser W (2002) Interactive rendering of large volume data sets. In: Proceedings of IEEE Visualization : 53–60

Hardisty F, Robinson AC (2011) The geoviz toolkit: using component-oriented coordination methods for geographic visualization and analysis. International Journal of Geographical Information Science 25:191–210

Hibbard WL, Anderson J, Foster I, Paul BE, Jacob R, Schafer C, Tyree MK (1996) Exploring coupled atmosphere-ocean models using Vis5D. International Journal of High Performance Computing Applications 10:211–222

Li J, Jiang Y, Yang C, Huang Q, Rice M (2013) Visualizing 3D/4D environmental data using many-core graphics processing units (GPUs) and multi-core central processing units (CPUs). Computers & Geosciences 59(9):78–89

MacEachren AM and Kraak M (2001) Research Challenges in Geovisualization. Cartography and Geographic Information Systems 28:3–12

McKaskle, GA, Rogers CC (2009) Integrated Data Viewer. US Patent 12/534,626 Aug.3 2009

Nickolls J, Buck I, Garland M, Skadron K (2008) Scalable Parallel Programming with CUDA. Queue 6(2) : 40–53

NVIDIA (2007) CUDA Compute Unified Device Architecture Programming Guide. NVIDIA Crop: Santa Clara, CA

Yang C, Wu H, Huang Q, Li Z, Li J (2011) Using spatial principles to optimize distributed computing for enabling the physical science discoveries. Proceedings of National Academy of Sciences of the United States of America 108(14):5498–5503

Chapter 7
A Parallel Algorithm to Solve Near-Shortest Path Problems on Raster Graphs

F. Antonio Medrano and Richard L. Church

Abstract The Near-Shortest Path (NSP) algorithm (Carlyle and Wood, Networks 46(2): 98–109, 2005; Medrano and Church, GeoTrans RP-01-12-01, UC Santa Barbara, 2012) has been identified as being effective at generating sets of good route alternatives for designing new infrastructure. While the algorithm itself is faster than other enumerative shortest path set approaches including the Kth-shortest path problem, the solution set size and computation time grow exponentially as the problem size or parameters increase, and requires the use of high-performance parallel computing to solve for real-world problems. We present a new breadth-first-search parallelization of the NSP algorithm. Computational results and future work for parallel efficiency improvements are discussed.

Keywords Parallel algorithms • Near-shortest paths • kth-shortest paths • Shortest path algorithms

7.1 Introduction

Since it was first declared in 1965, Moore's Law has correctly predicted that the number of transistors on an integrated circuit, and thus computational power, would double every 2 years. Up until 2005, the additional transistors allowed both processor clock speeds to increase and more advanced instruction-level parallelism, which resulted in overall computational processing power of single-threaded code to

F.A. Medrano (✉) • R.L. Church
Department of Geography, University of California at Santa Barbara,
Santa Barbara, CA, USA
e-mail: medrano@geog.ucsb.edu; church@geog.ucsb.edu

X. Shi et al. (eds.), *Modern Accelerator Technologies for Geographic Information Science*, 83
DOI 10.1007/978-1-4614-8745-6_7, © Springer Science+Business Media New York 2013

follow Moore's Law. But around 2005, heat dissipation became a limitation on further practical increases in processor clock speeds, and instead processor makers began using higher transistor densities to pack multiple computer processors onto a single chip, known as multi-core processors.

Innovation in multi-core processors, starting with dual-core, then quad-core, to now in 2013 where some processors have 8-cores capable of simultaneously handling 16-threads, have allowed the progress of Moore's Law to continue. But as scientists and programmers run programs on larger and more complicated data sets, they can no longer rely on simply higher processor clock speeds to improve the performance of their codes (Sutter 2005). Instead, to continue to reap the benefits of Moore's Law, programmers must now write their programs to take advantage of multi-core processors and increasingly inexpensive parallel computing clusters. This requires looking at ways to incorporate concurrency and multi-threading into their codes, so that independent control flows can be distributed over numerous processors. Some algorithms are easier to "parallelize" than others; for example, a Monte Carlo simulation entails running a model numerous times with various different initial conditions as input. Since each model simulation is an independent calculation to every other model simulation, then individual computations can simply be assigned to separate processors without the need for any communication between the processors while computing. Unfortunately, most programs are not so simple to make parallel, and more likely a programmer will have to split data and tasks into numerous pieces, perform some distributed concurrent partial computation, communicate intermediate results between various processors and then define new task and data pieces for further partial computation, continuing until the computation is complete; a sort of wash, rinse, and repeat. Communication speeds between processors become a performance bottle-neck, trade-offs between fine-grained and coarse-grained parallelism must be calibrated for running on different hardware configurations, and race conditions and deadlocks open up a whole new Pandora's box of software bugs that must be resolved.

But aside from the nuts and bolts associated with parallel programming, some algorithms are fundamentally difficult to split into concurrent processes. For example, parallel irregular graph traversal algorithms remain an active area of research, as these are inherently difficult to code (Bader et al. 2008; Cong et al. 2008; Chhugani et al. 2012; Merrill et al. 2012). This chapter addresses an example of one such irregular graph traversal: the parallelization of a depth-first-search (DFS) path algorithm that has been proven to be inherently sequential (Reif 1985), and thus difficult to implement in parallel. While the approach described here does make some progress in being able to make this path algorithm concurrent, our results suggest that there is considerable room for improvement, and we suggest future research plans and their associated challenges toward the end of this chapter. While this is a difficult problem, the development of new parallel path algorithm approaches is an important need when facing complex tasks such as robot operations and corridor planning on increasingly large and more complex networks.

Single objective shortest or least cost path tools are common in present day GIS software packages. These tools are useful in computing an optimal route over a terrain or road network, in which some objective such as distance or cost is to be minimized. Real-world problems are often more complicated than simply minimizing a single objective, and thus designers often need to solve more complex path problems such as the resource constrained shortest path (minimize cost A while not exceeding some quantity of cost B) or the multi-objective shortest path (minimize a weighted sum of numerous costs). The constrained shortest path and related problems are NP-Complete, and are thus quite difficult to solve.

Various methods using enumeration algorithms have been published for solving the resource constrained shortest path (Beasley and Christofides 1989; Carlyle et al. 2008; Handler and Zang 1980) and multi-objective shortest paths (Clímaco and Coutinho-Rodrigues 1988; Coutinho-Rodrigues et al. 1999; Raith and Ehrgott 2009). Most of these methods involve enumeration of some sort, either solving a Kth Shortest Path (KSP) problem, which returns a ranked list of shortest paths from an origin to a destination; or solving a Near-Shortest Path (NSP) problem, which returns a set of paths from an origin to a destination such that all are less than some defined cost. For loopless paths, algorithms for solving the NSP problem are much more efficient than those for solving KSP problems, and it has been shown that it is faster to solve the KSP by first solving an NSP and then post-processing the output (Carlyle and Wood 2005). This being the case, our research focuses on parallelizing the fastest known NSP algorithm (Carlyle and Wood 2005). Parallelization is necessary to solve large-scale problems, since we show that the solution set size will grow exponentially as problem sizes increases.

7.2 Background

Shortest path algorithms defined for network problems have been an active area of computational research since the 1950s. While shortest path routing has been a fundamental human problem since the dawn of time, Alex Orden in 1956 (Orden 1956) was the first to formulate mathematically the shortest path problem, which he did as a linear program. This was followed shortly by a number of different algorithms developed to solve the shortest path problem, including the well-known Bellman-Ford Algorithm (Bellman 1958) and Dijkstra's Algorithm (Dijkstra 1959). Since then, there have been many advances and refinements in shortest path algorithms, and recent comparisons between various methods can be found in Cherkassky et al. (1996), Zeng and Church (2009), and Zhan and Noon (2000).

The Kth-Shortest Path (KSP) Problem is an extension of the shortest path problem on a network, where the goal is to return the 1st, 2nd, 3rd, …, Kth shortest paths that exist between a pre-specified origin and destination. Initially formulated by Bock, Kantner, and Haynes (1957), good algorithms for solving this problem for loopless paths have been developed by Hoffman and Pavley (1959), Yen (1971), and Katoh et al. (1982). A complete literature review can be found in Medrano and Church (2011).

The Near-Shortest Path (NSP) Problem, originally formulated by Byers and Waterman (Byers and Waterman 1984), is a slight variation of the KSP problem. Unlike the KSP, which returns a ranked list of the k shortest paths on the network, the NSP problem returns all distinct paths on the network between an origin and destination longer than the shortest path within a prescribed threshold, ε, expressed as a decimal fraction. If the shortest path has length L_{sp}, then the NSP returns all paths of length $\leq (1+\varepsilon) \times L_{sp}$.

7.3 Carlyle and Wood's Near-Shortest Path Algorithm

In their 2005 paper, Carlyle and Wood (2005) present two different algorithms for finding loopless NSPs, ANSPR0 and ANSPR1 (Algorithm Near Shortest Paths Restricted 0 and 1 respectively). ANSPR0 is based on the Byers and Waterman (1984) method with a modification to output only loopless paths. The general idea is to find all paths of length $\leq D$ on the network, where $D = (1+\varepsilon) \times L_{sp}$. First, it solves the reverse shortest path tree (all shortest paths from the destination to all other nodes on the network) to acquire the shortest path cost from any node to the destination, t. This is the only time that a traditional shortest path algorithm is used. It then uses depth-first search (DFS) and a first-in last-out stack to generate the set of NSP's. This approach is very efficient because the DFS uses only fast addition/comparison operations, and never has to repeat any shortest path calculations in the process of generating paths. While it has an exponential worst-case complexity, it takes a pathological example to create such behavior. ANSPR1, has a better worst-case complexity, but when implemented was shown to run slower than ANSPR0. Combined with a binary search tree, Carlyle and Wood showed that the ANSPR1 algorithm could be modified to solve the KSP problem much faster than the fastest loopless KSP method of Katoh et al. (1982) as implemented by Hadjiconstantinou and Christofides (1999). While no other experiments have been published that compare Carlyle and Wood's algorithm to other KSP algorithms, Carlyle et al. (2008) argue that enumerating paths in order of length requires undue computational effort, and if it is not necessary to use KSP then the NSP is far superior.

7.4 The Need for Parallelization

To demonstrate the need for a parallelized approach to the NSP algorithm, we first wrote a serial JAVA implementation of the NSP algorithm and tested it on two networks:

1. 20×20 manually fabricated raster. This network contains 400 nodes and 2,850 undirected arcs.
2. 80×80 subset of the Maryland Automated Geographic Information System (MAGI) database. This network contains 6,400 nodes and 49,770 undirected arcs.

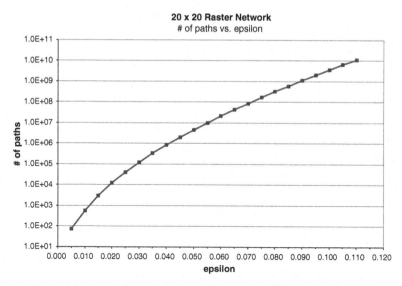

Fig. 7.1 20×20 network, log number of paths generated by the ANSPR0 vs. epsilon

Both of these data sets were first used by Huber and Church (1985). Both networks have undirected "queen's and knight's move" arcs emanating from each node, as this raster network model has been shown to offer a good compromise between accuracy and computational burden (Huber and Church 1985).

To characterize the solution set growth rate, we plotted the number of paths output by NSP for numerous values of ε on both networks on a log-linear plot. Figure 7.1 shows this plot for the 20×20 network. The logarithmic y-axis is of the number of paths output, and the x-axis is the ε value. The result is a straight-line trend, indicating an exponential growth in the number of paths as ε increases. Similarly, as seen in Fig. 7.2, the computation time as a function of ε was found to be proportional to solution size growth, and when plotted on the log-linear axis also showed exponential growth with respect to ε.

The NSP algorithm generated nearly 4 billion solutions on the 20×20 raster region when the epsilon value was set at 0.10, i.e. within 10 % of the optimal path. Because the number of paths for each given value of ε is much higher for the 80×80 network as compared to the 20×20, the range of epsilons used in our 80×80 experiments were an order of magnitude smaller than those in our 20×20 experiments. For example, for $\varepsilon = 0.005$, on the 20×20 data this generated a solution set of 73 near-shortest paths; but for the 80×80 with the same $\varepsilon = 0.005$, there were 510,343,616 such paths.

These experiments demonstrated that generating a set of NSPs can be an enormous task and may overwhelm computational resources as one increases the network size or increases the value of ε. Generating all paths within 0.75 % of the shortest path length on the 80×80 network took more than 4 days using an Intel

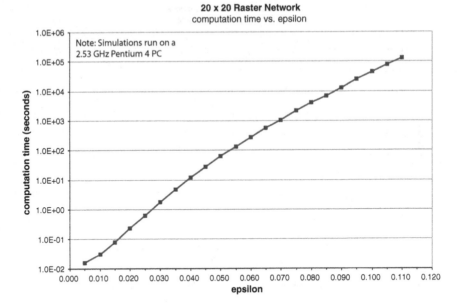

Fig. 7.2 20×20 network, log computation time of ANSPR0 vs. epsilon

Core i7 desktop and employing a serial JAVA implementation, even though this application was able to generate 185,000 paths per second! The quad-core processor used was capable of running two threads per core, yet the serial code was using only one out of the eight potential threads of this processor. While we make no claims that the serial code cannot be further optimized, the reality is that even with the best serial code, generating all paths within 10 % of the shortest path on a 100 megapixel raster is beyond the reach of any commercial off-the-shelf computer running a serial NSP code. Instead, in order to have any hope of making this computation, one must take advantage of the full parallel capabilities of modern processors.

7.5   Parallelizing Depth-First-Search

Our proposed technique of converting the Carlyle and Wood depth-first-search NSP algorithm into a parallel algorithm begins with performing an initial breadth-first-search (BFS) generation of all NSPs emanating from the origin point. Rather than completely building one NSP at a time with DFS, the BFS simultaneously builds the start of all NSPs. BFS naturally forms a tree structure, with concurrent paths sharing branches until they deviate, and the end nodes of the various path stems resulting as different leaves on the tree. We stored the BFS paths in a "*trie*" data structure (Hadjiconstantinou and Christofides 1999; Aho et al. 1983), which is an efficient data structure for storing sets of paths which partially overlap. If the entire computation were performed as a BFS, then every leaf of the final *trie* would be the destination node of the network. In our method, the BFS runs until there are as many leaves on the BFS tree as there are processors available for computation, at which point each processor is then tasked with running the traditional DFS NSP algorithm using a *trie* leaf as a starting point, finding all paths from that point which are less than the threshold length minus the path-length from the origin to the leaf starting node.

7.6   Analysis of Implementation

Prototyping was done using UCSD's Triton Supercomputer. Triton consists of 256 gB222X Appro blade nodes, each containing 2 quad-core Intel Nehalem 2.4 GHz processors, 24 GB of memory, and is capable of a peak processing power of 20 TeraFlops. Our code was written in C++ using the MPI extension to communicate between the different processors/nodes.

Table 7.1 contains data collected from various runs of our first generation parallel code implementation on the 20×20 data set on the Triton nodes. The columns show epsilon value, number of processors, total NSP runtime in seconds, total paths found, paths found on the leaf of fewest paths (Min Paths Leaf), paths found on the leaf of most paths (Max Paths Leaf), speedup, and parallel efficiency. When using multiple processors, the closer the value of parallel efficiency is to 1.00 the better. By definition, when one uses only one processor, it will be rated at 100 % efficiency for that one processor. The main objective in parallelizing a routine is to hopefully use all processors efficiently with no idle time and reach a parallel efficiency of 1.0 overall, although this rarely happens for all but the most trivial algorithms.

In the computational results shown in Table 7.1 we see that good parallel efficiency was achieved when the ratio between the maximum number of paths found on a leaf and the minimum number of paths found on a leaf is not too great. For example, when $\varepsilon = 0.05$, and 5 processors were employed (BFS depth = 1), the ratio was approximately 6:1 "max to min paths". This resulted in a very respectable 0.61 value of parallel efficiency. With $\varepsilon = 0.05$ and employing 48 processes (BFS depth = 2), the "max to min paths" ratio was approximately 50,000:1. The minimum path leaf quickly finished its work in 0.2 ms, while the maximum path leaf took 2.5 s to complete. This significant amount of relative idle time resulted in a lesser parallel efficiency of 0.18.

Table 7.2 gives results for computational tests on the 80×80 data using the same parallel implementation. This experiment produced even larger discrepancies between the number of paths found in the "max" sized leaf and the number of paths found in the "min" sized leaf, resulting in an unimpressive speedup of 1.61 when using 38 processors, which is equivalent to a parallel efficiency of 0.04.

As a result of the independent computation of each leaf, we found that that the expected overall parallel efficiency followed this relationship:

Table 7.1 Parallel NSP runtime results on 20×20 network

Epsilon	Number of processors	Time (s)	Total paths		Min paths leaf	Max paths leaf	Speed-up	Parallel efficiency
0.05	1	21.73	4,601,053	Paths	4,601,053		1.00	1.00
				Time	21.729			
				Paths/s	211,747			
0.05	5	7.07	4,601,053	Paths	247,446	1,530,887	3.07	0.61
				Time	1.30064	7.07048		
				Paths/s	190,249	216,518		
0.05	48	2.51	4,601,053	Paths	11	565,901	8.67	0.18
				Time	0.000195	2.50676		
				Paths/s	56,403	225,750		
0.07	1	392.37	86,384,393	Paths	86,384,393		1.00	1.00
				Time	392.373			
				Paths/s	220,159			
0.07	5	119.94	86,384,393	Paths	5,782,131	26,620,106	3.27	0.65
				Time	27.3463	119.939		
				Paths/s	211,441	221,947		
0.07	50	34.39	86,384,393	Paths	137	8,129,092	11.41	0.23
				Time	0.00161	34.3939		
				Paths/s	85,091	236,353		

Table 7.2 Parallel NSP runtime results on the 80×80 network

Epsilon	Number of processors	Time (s)	Total paths		Min leaf path	Max leaf paths	Speed-up	Parallel efficiency
0.003	1	33.21	4,459,050	Paths	4,459,050		1.00	1.00
				Time	33.21			
				Paths/s	134,253			
0.003	5	25.51	4,459,050	Paths	3,462	3,475,928	1.30	0.26
				Time	0.03324	25.5127		
				Paths/s	104,152	136,243		
0.003	11	25.49	4,459,050	Paths	852	3,472,466	1.30	0.12
				Time	0.01286	25.4856		
				Paths/s	66,251	136,252		
0.003	12	25.50	4,459,050	Paths	852	3,472,466	1.30	0.11
				Time	0.01262	25.5003		
				Paths/s	67,501	136,174		
0.003	14	24.29	4,459,050	Paths	600	3,462,254	1.37	0.10
				Time	0.00817	24.2906		
				Paths/s	73,475	142,535		
0.003	22	21.95	4,459,050	Paths	300	3,175,358	1.51	0.07
				Time	0.00408	21.9474		
				Paths/s	73,511	144,680		
0.003	38	20.68	4,459,050	Paths	216	3,033,530	1.61	0.04
				Time	0.00527	20.68		
				Paths/s	41,018	146,714		

$$parallel\ efficiency = \frac{Paths_{Total}}{p \times Paths_{Max}}$$

where $Paths_{Total}$ is the total number of paths found for the given input parameters and data, and $Paths_{Max}$ is the maximum number of paths found by one processor, and p is the number of processors. Additionally, as $Paths_{Max} \rightarrow Paths_{Total}/p$, then *parallel efficiency* $\rightarrow 1$. This is essentially an example of Amdahl's Law (Amdahl 1967) in action, which states that the potential parallelism available in any program is limited by the amount of work that must be run sequentially. This points towards the need to distribute the work more evenly in order to make the most efficient use of all processors.

7.7 Implementation Challenges: Distributing Workload

The load imbalances in this problem come from performing a depth-first search on a raster network, where the task workload sizes are completely unknown until after execution is completed. Therefore, offline partitioning or scheduling algorithms cannot be used beforehand, as there is not enough information available in order to make use of such schemes. The following is a description of several methods for load balancing and how well they could apply to our parallel approach to the NSP problem. For further details, please refer to Medrano and Church (2012).

Randomized Task Distribution. As it stood before, the code distributed the work by running a BFS algorithm until the tree had as many leaves as there were processors, then assigned one leaf to each processor for it to run to completion. The drawback was that some leaves contained far more work than others, resulting in lots of idle processor time for some of the processors.

A randomized task distribution approach would be based on generating far more leaves than processors, then assign these tasks randomly to each processor. By randomly distributing sufficient work chunks of unknown size to numerous processors, the hope is that overall work for each processor averages out to be somewhat similar. Adler et al. (Adler et al. 1995) show that when using randomized algorithms on normal or Poisson distributions of workload, in order to get a "good" balance one must generate at the very least $p \log p$ tasks, where p is the number of processors. In a worst-case-scenario though, a large outlier could still result in an overall work imbalance

Dynamic Centralized Scheduling. Centralized scheduling uses an as-needed approach for assigning tasks. Like randomized task distribution, centralized scheduling first generates a list of tasks ($\gg p$), then assigns the first p tasks to the various processors to compute. When a processor completes a task, it asks the scheduler for another task. The scheduler assigns a new task, removes it from the list, and this process would continue until all tasks have been assigned and computed. This method is susceptible to the possibility of an abnormally large task being assigned last.

Dynamic Work Stealing. Dynamic work stealing is an approach that assigns all work to all processors at the start; then when one processor completes its tasks, it steals part of a task from another processor in order to have more work to do. This approach is certainly viable for a DFS algorithm; and if one does not consider communication time between processors, it has the possibility of producing the best theoretical results. Unfortunately, it is also far more difficult to implement than any of the other options. Even if implemented, one has to select a strategy for selecting which processor to steal work from, including asynchronous round robin, global round robin, and random polling/stealing. It has been proven that a random polling/stealing approach is theoretically just as effective as the other two approaches (Blumofe and Leiserson 1994), although local communication priority is preferred in practice. This application could use a worker queue that at each time assigns work from the processor at the front of the queue. Any time a processor steals work or gets stolen from, it then gets placed at the back of the queue, ensuring it won't get stolen from immediately afterward, essentially a FIFO scheme.

Workload Prediction. One reason why it is difficult to balance the workload on depth-first-search irregular graph traversal algorithms is because the amount of work in each branch varies widely, and is unknown beforehand. We have considered working to identify heuristics that estimate the amount of work needed to resolve each leaf of the BFS tree. If effective, this would allow one to fathom/trim the tree in portions that have low expected work, while continuing to split leaves on portions with higher expected work. This would hopefully result in work chunks of more uniform size and avoid the inefficiencies caused by massively disparate work task sizes.

For the parallelized NSP algorithm, it appears that the first two approaches would suffer from the possibility of large work chunks superseding the benefits of random prescheduling or dynamic work scheduling. On our 80×80 data set (Table 7.2), even dividing the work into 38 chunks, the maximum sized still accounted for 68 % of the total paths. This is far from the Gaussian or Poisson distribution that is necessary for random prescheduling to be effective. Dynamic work stealing has no theoretical drawbacks if implemented properly, but is exceedingly difficult to program for graph problems. In looking for an optimal balance between performance gains and ease of implementation, workload prediction heuristics for the purpose of developing the BFS tree only in portions with a high-expected workload seem most promising in being an efficient method for more evenly distributing the workload across processors.

Additionally, the best way to use the strengths and hide the weaknesses of any approach is to combine it with another complimentary approach. For example, a hybrid workload prediction/work stealing approach could show promise in giving a relatively even initial work distribution, then leveling task loads towards the end using dynamic work stealing. Any hybrid approach would be the most difficult to implement, as it requires developing several approaches, as well as cooperatively integrating them together.

7.8 Conclusions

The goal of this research was to develop an efficient parallel implementation of the fastest NSP algorithm. The approach described here uses breadth-first-search to split the work up in a pleasingly parallel fashion, and was able to show a significant speedup when computing with multiple processors. Large variances in the work-chunk sizes though prevented this approach from running at theoretically optimal parallel efficiencies. Further work is needed in exploring approaches to more evenly spread workload across various processors. We described possible methods for further improvements in parallel efficiency, and of those we recommend devising a predictive metric that could be used to estimate the work on each portion of the BFS tree, and stunting tree growth where small work would be expected, followed by a work-stealing scheme during the computation. We expect that the former would result in a more consistent set of work-chunks, while the latter would even-out any remaining imbalances, leading to improved overall efficiency and performance of the parallel code. Further research would aim to develop an effective GIS tool able to solve larger and more complicated path routing problems.

Acknowledgements We would like to thank the Environmental Sciences Division of Argonne National Laboratories for providing the funding to conduct this research (1F-32422).

References

Adler, M., et al.: Parallel randomized load balancing. In: Proceedings of the twenty-seventh annual ACM symposium on Theory of computing, 238–247 ACM City (1995)

Aho, A.V., J.E. Hopcroft, and J. Ullman: Data structures and algorithms. Addison-Wesley Longman Publishing Co., Inc. Boston, MA, USA, (1983)

Amdahl, G.M.: Validity of the single processor approach to achieving large scale computing capabilities. In: AFIPS. Atlantic City, N.J.: ACM. (1967)

Bader, D.: Petascale computing for large-scale graph problems. Parallel Processing and Applied Mathematics, 166-169 (2008)

Beasley, J.E. and N. Christofides: An Algorithm for the Resource Constrained Shortest-Path Problem. Networks, 19(4), 379–394 (1989)

Bellman, R.E.: On a routing problem. Q. Applied Math, 1687–90 (1958)

Blumofe, R.D. and C.E. Leiserson: Scheduling multithreaded computations by work stealing. In: Foundations of Computer Science, 1994 Proceedings., 35th Annual Symposium on. IEEE. (1994)

Bock, F., H. Kantner, and J. Haynes: An algorithm (the r-th best path algorithm) for finding and ranking paths through a network. Research report, Armour Research Foundation of Illinois Institute of Technology, Chicago, Illinois, (1957)

Byers, T. and M. Waterman: Determining all optimal and near-optimal solutions when solving shortest path problems by dynamic programming. Operations Research, 32(6), 1381–1384 (1984)

Carlyle, W.M. and R.K. Wood: Near-shortest and K-shortest simple paths. Networks, 46(2), 98–109 (2005)

Carlyle, W.M., J.O. Royset, and R.K. Wood: Lagrangian Relaxation and Enumeration for Solving Constrained Shortest-Path Problems. Networks, 52(4), 256–270 (2008)

Chhugani, J., et al.: Fast and Efficient Graph Traversal Algorithm for CPUs: Maximizing Single-Node Efficiency. In: Parallel & Distributed Processing Symposium (IPDPS), 2012 IEEE 26th International. IEEE. (2012)

Cherkassky, B.V., A.V. Goldberg, and T. Radzik: Shortest paths algorithms: theory and experimental evaluation. Mathematical programming, 73(2), 129–174 (1996)

Clímaco, J. and J. Coutinho-Rodrigues: On an interactive bicriteria shortest path algorithm. Lisbon, Portugal. (1988)

Cong, G., et al.: Solving large, irregular graph problems using adaptive work-stealing. In: Parallel Processing, 2008. ICPP'08. 37th International Conference on. IEEE. (2008)

Coutinho-Rodrigues, J., J. Climaco, and J. Current: An interactive bi-objective shortest path approach: searching for unsupported nondominated solutions. Computers & Operations Research, 26(8), 789–798 (1999)

Dijkstra, E.W.: A note on two problems in connexion with graphs. Numerische Mathematik, 1(1), 269–271 (1959)

Hadjiconstantinou, E. and N. Christofides: An efficient implementation of an algorithm for finding K shortest simple paths. Networks, 34(2), 88–101 (1999)

Handler, G.Y. and I. Zang: A dual algorithm for the constrained shortest path problem. Networks, 10(4), 293–309 (1980)

Hoffman, W. and R. Pavley: A Method for the Solution of the N th Best Path Problem. Journal of the ACM (JACM), 6(4), 506–514 (1959)

Huber, D.L. and R.L. Church: Transmission Corridor Location Modeling. Journal of Transportation Engineering-Asce, 111(2), 114–130 (1985)

Katoh, N., T. Ibaraki, and H. Mine: An efficient algorithm for k shortest simple paths. Networks, 12(4), 411–427 (1982)

Medrano, F.A. and R.L. Church: A New Parallel Algorithm to Solve the Near-Shortest-Path Problem on Raster Graphs. GeoTrans RP-01-12-01, UC Santa Barbara (2012)

Medrano, F.A. and R.L. Church: Transmission Corridor Location: Multi-Path Alternative Generation Using the K-Shortest Path Method. GeoTrans RP-01-11-01, UC Santa Barbara (2011)

Merrill, D., M. Garland, and A. Grimshaw: Scalable GPU graph traversal. In: Proceedings of the 17th ACM SIGPLAN symposium on Principles and Practice of Parallel Programming. ACM. (2012)

Raith, A. and M. Ehrgott: A comparison of solution strategies for biobjective shortest path problems. Computers & Operations Research, 36(4), 1299–1331 (2009)

Reif, J.H.: Depth-first search is inherently sequential. Information Processing Letters, 20(5), 229–234 (1985)

Sutter, H.: The free lunch is over: A fundamental turn toward concurrency in software. Dr. Dobb's Journal, 30(3), 202–210 (2005)

Orden, A.: The transhipment problem. Management Science, 2(3), 276–285 (1956)

Yen, J.Y.: Finding the K Shortest Loopless Paths in a Network. Management Science, 17(11), 712–716 (1971)

Zeng, W. and R.L. Church: Finding shortest paths on real road networks: the case for A*. International Journal of Geographical Information Science, 23(4), 531–543 (2009)

Zhan, F. and C. Noon: A Comparison Between Label-Setting and Label-Correcting Algorithms for Computing One-to-One Shortest Paths. Journal of Geographic Information and Decision Analysis, 4(2), 1–11 (2000)

Chapter 8
CUDA-Accelerated HD-ODETLAP: Lossy High Dimensional Gridded Data Compression

W. Randolph Franklin, You Li, Tsz-Yam Lau, and Peter Fox

Abstract We present *High-dimensional Overdetermined Laplacian Partial Differential Equations* (HD-ODETLAP), an algorithm and implementation for lossy compression of high-dimensional arrays of data. HD-ODETLAP exploits autocorrelations in the data in any dimension. It also adapts to regions in the data with varying value ranges, resulting in the maximum error being closer to the RMS error. HD-ODETLAP compresses a data array by iteratively selecting a representative set of points from the array. That set of points, efficiently coded, is the compressed dataset. The compressed dataset is uncompressed by solving an overdetermined sparse system of linear equations for an approximation to the original array. HD-ODETLAP uses NVIDIA CUDA called from MATLAB to exploit GPU parallel processing to achieve considerable speedup compared to execution on a CPU. In addition, HD-ODETLAP compresses much better than JPEG2000 and 3D-SPIHT, when fixing either the average or the maximum error. An application is to facilitate storage and transmission of voluminous datasets for better climatological and environmental analysis and prediction.

Keywords ODETLAP • Lossy compression • Geospatial • High dimensional data • NVidia GPU

8.1 Introduction and Background

The research theme of this paper is the use of modern accelerator technologies to address the problem of storing increasing volumes of multidimensional geospatial data using lossy compression. The good compression algorithms needed to

W.R. Franklin (✉) • Y. Li • T.-Y. Lau • P. Fox
Rensselaer Polytechnic Institute, Troy, NY, USA
e-mail: mail@wrfranklin.org; liyou.rpi@gmail.com; rpi.laut@gmail.com; pfox@cs.rpi.edu

X. Shi et al. (eds.), *Modern Accelerator Technologies for Geographic Information Science*, 95
DOI 10.1007/978-1-4614-8745-6_8, © Springer Science+Business Media New York 2013

maximize the feasible dataset size are quite compute-intensive. Also, the datasets' multi-dimensional structure is not exploited by current compression algorithms. In this paper, we introduce **HD-ODETLAP**, a *High-Dimensional Over-determined Laplacian Partial Differential Equation* compression algorithm and implementation. HD-ODETLAP has various versions, such as 4D-ODETLAP and 5D-ODETLAP. Their novel technique resides in expressing the problem as an overdetermined sparse system of linear equations.

Our test datasets are the *World Ocean Atlas 2005* and *2009* (Locarnini et al. 2010) from the National Oceanographic Data Center (NODC) and the National Geophysical Data Center (NGDC), subsidiaries of the NOAA Atmospheric Administration (NOAA). They contain marine properties *temperature, salinity, nitrate*, and *silicate* over 12 months at 24 standard depths in the ocean. They can be considered to be five-dimensional: *(latitude, longitude, depth, time, property value)*, of size 180 ×360 ×24 ×12 ×4. The data is autocorrelated along each dimension: small changes in any coordinate cause small changes in the measured quantity. This applies even to the fifth dimension; the observed variables do, to some extent, vary in concert.

However, current compression methods treating the data as a single-dimensional stream of bytes ignore that correlation. To the extent that HD-ODETLAP can exploit this property, HD-ODETLAP will compress better, and it does on the test data. HD-ODETLAP also adapts to nonhomegenelities in the data, where some local regions may have larger value ranges than other regions. Therefore, an HD-ODETLAP representation with a given RMS error may have a smaller maximum error than do other methods. This is also confirmed by experiment. However, the problem with HD-ODETLAP is its compute-intensiveness, and so, modern accelerator techniques are desirable.

ODETLAP, initially in a 2D version, was developed as part of a project to lossily compress terrain (elevation of the earth's surface above the geoid or assumed sea level). A goal was to facilitate operations such as multi-observer siting, and then path planning to avoid the observers (the *Smugglers and Border Guards Problem*). The initial goal was to interpolate terrain from isolated data points and contour lines. The contour lines may be kidney-bean shaped and might have gaps. Those properties caused problems with earlier interpolation algorithms, such as running straight lines in the eight cardinal directions from the test point to the nearest contour line. In addition, terrain data often have limited precision, and may be mutually inconsistent. However, the interpolated slope should be continuous across lines of data points; that is, a line of data should not cause a visible "kink" in the generated surface. ODETLAP addresses these problems well (Franklin 2011), (Franklin et al. 2006), (Inanc 2008), (Li 2011), (Muckell 2008), (Stookey 2008), (Stookey et al. 2008), (Tracy 2009), (Xie 2008). Figure 8.1 shows how successfully ODETLAP handles an example designed to be very difficult, with input contours with sharp corners. Nevertheless the interpolated output surface has smooth silhouette edges, and an inferred peak inside the innermost contour. All this is achieved with a mean error of 2.7 % of the data range (computed on the known points) and a maximum error of 12.9 %.

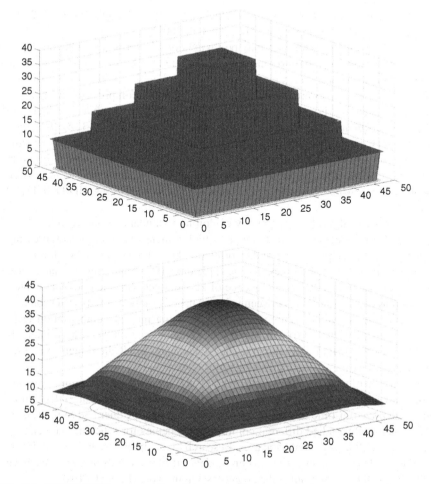

Fig. 8.1 ODETLAP fitting a smooth surface to nested square contours

8.2   Compression Basics

Various compression methods have been created for lower dimensional data, such as 3D image sequences (Menegaz and Thiran 2002) and 4D (for example, 3D spatial + temporal) functional magnetic resonance imaging (FMRI) (Lalgudi et al. 2005). Among those methods, the wavelet-based lossy and lossless ones are the most popular. For example, JPEG 2000 (Taubman et al. 2002) uses irreversible and reversible wavelet transform for its lossy and lossless compression for 2D images. For 3D video data, *Three Dimensional Set Partitioning in Hierarchical Trees* (3D-SPIHT) (Kim and Pearlman 1997) also bases its compression scheme on 3D wavelet transforms. There are some 4D compression algorithms, such as 4D wavelets (Yang et al. 2006), run length encoding (RLE) (Anagnostou et al. 2000) and discrete cosine transform (DCT) (Lum et al. 2001).

Data compression techniques may be *lossless* or *lossy*. Lossless schemes allow exact reconstruction of the original data but suffer from a low compression ratio. Lossy schemes produce much more compact datasets, at an, often modest, increase, δ, in the dataset's RMS error, ε. If $\delta << \varepsilon$, then the lossy compression has not cost very much accuracy. For example, suppose that a uniformly distributed temperature parameter is quantized to integral degrees, so that $\varepsilon = 1/(2\sqrt{3}) \approx 0.3$. $\delta = 0.1$ might allow a much smaller dataset with a small cost in increased RMS error. Experiments on elevation data that trade off error with compressed size are described in Franklin and Said (1996). The reader might also experiment by comparing compression quality and size when generating JPEG images. For these reasons, HD-ODETLAP is lossy.

Compression algorithms operate by exploiting redundancies and correlations in the data. Current techniques that compress high dimensional data by compressing 2D or 3D slices separately ignore the correlation between the slices. We don't.

3D compression methods can also be applied on 4D spatial-temporal data, since 4D data could be handled as a sequence of 3D volumes. But methods that exploit the temporal autocorrelation between volumes usually outperform their 3D counterparts, such as the video compression methods using motion compensation technique (Sanchez et al. 2006) and 4D-SPIHT (Ziegler et al. 2004). These methods fully utilize the spatial and temporal data redundancy in all the dimensions; thus they can achieve a higher compression ratio.

In GIS, there is an increasing awareness of the significance of compression and progressive transmission of high dimensional spatial data. Plaza et al. (2010) introduces adaptive run-time data compression of spatial data. Kidner and Smith (2003) comprehensively studied different compression schemes for efficient maintenance and dissemination of spatial databases. Research on compression and spatial decorrelation has been done in digital terrain models, (Bjøke and Nilsen 2002). We have published a 3D oceanographic data compression method, (Li et al. 2010).

8.3   ODETLAP Definition

We will first present a 2D version of *Five Dimensional Over-Determined Laplacian Partial Differential Equations* (5D-ODETLAP), then extend it. The domain is a 2D array of elevations. The (i, j) entry has the position (x, y) when projected vertically onto the geoid. (x, y) are typically either latitude and longitude, or Universal Transverse Mercator (UTM) coordinates. z is the elevation above the geoid.

ODETLAP has two components. The smaller component is to interpolate from a set of known point elevations to a complete elevation array. The larger component is to loosely compress an elevation array by selecting an important set of points S from the array. S is the compressed representation. S is uncompressed by using ODETLAP to interpolate back to the whole elevation array. The process is illustrated in Fig. 8.2.

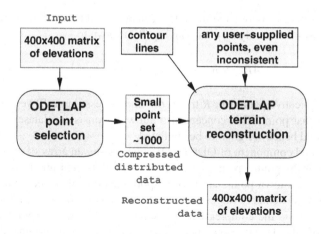

Fig. 8.2 ODETLAP process

The interpolation component of 2D-ODETLAP is an extension of the Laplacian PDE

$$\frac{\delta^2 z}{\delta x^2} + \frac{\delta^2 z}{\delta y^2} = 0 \tag{8.1}$$

to an overdetermined linear system. The purpose of this system is to interpolate elevations from a small set of known points to the entire array. This system comprises two types of equations. First, every nonborder point, known or unknown, induces an averaging equation that sets its value to the average of its neighbors:

$$4z_{ij} = z_{i-1,j} + z_{i+1,j} + z_{i,j-1} + z_{i,j+1} \tag{8.2}$$

Second, we add another equation for each known point,

$$z_i = s_i \tag{8.3}$$

where s_i stands for the input known value of the point, and z_i its output computed value. Since ODETLAP is lossy, these values will differ slightly; the fitted surface does not exactly interpolate the known data. This system is over-determined, since there are more equations than unknowns. Being over-determined, it has essentially different mathematical properties from a Laplacian, in spite of the superficial similarities. For example, with ODETLAP, unlike with a Laplacian, interior local extrema can be inferred. Fitting a nested set of contours produces a rounded hill, not a flat topped mesa. Also, unlike with a Laplacian PDE, the inferred surface is more continuous across lines of known points, and so, the data points are not as visible (or not at all visible) in the generated surface, as shown in Fig. 8.1.

The least squares solution to this system can be biased by weighting the different equation classes differently by changing Eq. (8.2) above to

$$4Rz_{ij} = R\left(z_{i-1,j} + z_{i+1,j} + z_{i,j-1} + z_{i,j+1}\right) \tag{8.4}$$

The multiplicative parameter R trades off smoothness (large R) versus accuracy (small R) at that point. This concept also allows inconsistent datasets of varying accuracies (and hence weights) to be conflated.

In the second component of ODETLAP, compressing an array of elevations, we select some of the points, perhaps 1%, as *known* points, and ignore the others. In some sense, the known points are more important. They might be mountain tops, valley bottoms, the ends of ridges, river confluences, etc. That set of known points, $S = \{(x_i, y_i, s_i)\}$, is the compressed representation of the data. Applying the first component of ODETLAP reconstructs an approximation to the original data.

There is no Eq. (8.4) for points on the border of the region. However those points' values are part of the equations of their nonborder neighbors. One might, wrongly, create an equation for a border point to set it equal to the average of the 2 or 3 neighbors that it does have. However that would impose an unwarranted bias towards horizontality on the surface at the border. To prevent the whole system from being underdetermined, there must be at least as many known points as border points. In practice, this is not a problem.

2D-ODETLAP is used to compress a surface, expressed as an array of elevations, as follows.

1. Select a random subset S of the points in the input array. Selecting 0.1% of the points is reasonable.
2. Use ODETLAP to compute an initial approximation of the dataset.
3. Of all the points in the array, find those that are farthest from that approximation. Again, 0.1% of the array is a reasonable heuristic.
4. If the (maximum, average, or whatever desired metric) error is adequate, then exit this loop. Otherwise:
5. Insert those worst points into S.
6. Go back to step 2.

The extension of ODETLAP from 2D to 3D or 4D is obvious. Each nonborder point now has 8 or 16, rather than 4, neighbors. In our current 5D-ODETLAP implementation, the fifth dimension in our dataset is a property, of a different nature than the other four, we currently implement a 4D-ODETLAP for each property (e.g., temperature, salinity, …) in the fifth dimension. Meanwhile the points' locations are constrained to be the same in each 4D-ODETLAP. This utilizes the fact that the locations of important points for one property are often also important points for the other properties. Therefore we need to encode the points' locations only once, saving considerable space. The representation of the compressed dataset is the set of positions and vector of values of the elements of S.

We encode the known point locations and the respective property values separately. Note that the bitmap of the positions is like a 2-value facsimile image. For both the set and vector, we tested various methods to optimize the coding. We used LEDA, a C++ class library for efficient data types and algorithms (Mehlhorn and Näher 1995), to build our own entropy coder. LEDA contains a wide range of simple coders, which could be easily combined; we built the following ten different coders.

- A0: Adaptive Arithmetic Coding
- BMRA: Burrows-Wheeler Transform + Move to Front + RLE for Runs of Zeros + Adaptive Arithmetic Coding
- BRA: Burrows-Wheeler Transform + RLE for Runs of Zeros + Adaptive Arithmetic Coding
- Huff: Adaptive Huffman Coding
- MRP: Move to Front + RLE for Runs of Zeros + Prediction by Partial Matching
- RMP: RLE for Runs of Zeros + Move to Front + Prediction by Partial Matching
- RDP: RLE for Runs of Zeros + Dictionary-based Coder + Prediction by Partial Matching
- PPM: Prediction by Partial Matching
- Dict_PPM: Dictionary-based Coder + Prediction by Partial Matching
- RLE_PPM: RLE for Runs of Zeros + Prediction by Partial Matching

8.4 Known-Point Position Compression

For the sparse binary 4D matrix representing the positions of the known points, we first used *Binary RLE* (binary run length encoding) to convert it into a integer vector, defined thus:

$$P = \{r_1 r_2 \ldots r_i \ldots r_k\} \tag{8.5}$$

where r_i represents the number of 0's before the i-th 1 in a row-major order. k is the total number of known points, i.e., the number of 1's.

Figure 8.3 shows the result of using the 17 different coders mentioned in the previous section to compress the 4D binary position file. The 10 coders from LEDA compress the integer vector from *Binary RLE*. We also include the result of the Bzip2 and JPEG 2000 coder for the integer vector. There are five possible coders in TIFF, which is widely used for compressing binary images: the International Telegraph and Telephone Consultative Committee (CCITT) Fax3, CCITTFax4, Lempel-Ziv-Welch (LZW), PackBits and Deflate. We present the TIFF compression results using them on a test dataset that is a 90 ×180 ×24 ×12 ×4 array from WOA 2009, with 91119 known points. The RLE_PPM coder is the winner, and so 5D-ODETLAP uses it.

The known-point position compression is lossless since otherwise the points' positions would move in an unpredictable way. Allowing a lossy compression here is a possible future research idea.

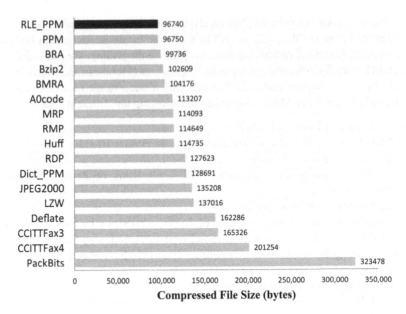

Fig. 8.3 Binary position file compressed file size by 17 different coders. The uncompressed 5D binary position file is 2332800 bytes

8.5 Known-Point Value Compression

For the vector of values for the known points, we used Lloyd relaxation to produce an optimal quantization for the floating point value. The best-known and earliest quantization algorithm, generalized Lloyd algorithm (GLA) Lloyd (1982) is based on minimizing the Mean Square Quantization error (MSE_q), which can be expressed as:

$$MSE_q = \sum_{i=1}^{N} \int_{d_i}^{d_{i+1}} \left(x - C(x)\right)^2 f_x(x)\,dx \qquad (8.6)$$

where x and $C(x)$ are input floating value and quantized output, respectively. N is the total number of reconstruction levels, and $f_x(x)$ is the probability density function (pdf) of the input vector. The quantization give us two files: one file contains the integer indices, while the other records floating point codebook.

After the quantization step, we apply a conditional delta encoding to the quantized integer indices. This part of the encoding is conditional because we apply it only when it benefits the compression. In particular, this step adds one sign bit to the integer vector. As shown in Table 8.1, this action usually improves the coding efficiency of the subsequent PPM coding only when adding such a bit allows us to store a value whose number of bits is a multiple of 4 or 8 (because the PPM coder acts on a byte stream).

Table 8.1 Influence of delta encoding on the size of the final compressed floating point value vector. The test data is the same as in Fig. 8.3

Integer range	W/o delta encoding (bytes)	W. delta encoding (bytes)	Change in file size (%)
0–8 (3 bits)	11,132	10,790	−3.170
0–16 (4 bits)	18,750	23,497	20.203
0–32 (5 bits)	32,388	32,882	1.502
0–64 (6 bits)	44,236	48,281	8.378
0–128 (7 bits)	58,596	50,916	−15.084
0–256 (8 bits)	59,712	72,132	17.218
0–512 (9 bits)	82,611	83,368	0.916
0–1024 (10 bits)	93,932	97,796	3.951

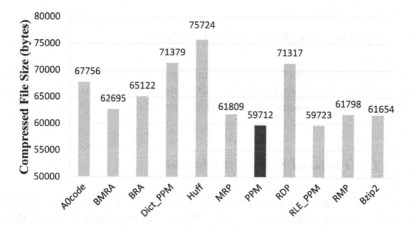

Fig. 8.4 Floating point value vector final compressed file size by 11 different coders

Figure 8.4 shows the result of using different coders to compress the quantized floating point vector without delta encoding. The input number of bits used for each value is 8. The test data is the same as in Fig. 8.3. We also include the result of the Bzip2 coder for comparison. This figure demonstrates that the PPM coder performs the best in our 5D-ODETLAP compression framework.

8.6   Challenge of Compute-Intensiveness

The biggest challenge of this compression framework is that solving overdetermined sparse linear systems is compute-intensive. The normal equations transformation, taking $Ax=b$ to $A^TAx=A^Tb$, reduces the computation time considerably. Indeed the system is now exactly determined, instead of overdetermined. However, this increases the memory requirement, and the computation is still slow.

For example, solving a 104976 ×104976 linear system takes up to 58 GB main memory and about 1.8 hours on a 2006-vintage workstation with four 2.4GHz processors and 60 GB of main memory running Ubuntu 10.04.2 and 64-bit MATLAB R2009a. Current processors might improve that by a small integer factor. The following sections show how modern GPU accelerator technologies give a large factor improvement.

A major sub-challenge is integrating various software tools described below into an efficient solution. Not having to reinvent those tools allows us to concentrate on our problem.

Those tools reduce the computation time so much that now the I/O time for transmitting data to the graphics processor is the dominant cost. So, the major future challenge will be to reduce that.

8.7 Application of Accelerator Technology

Modern accelerator technology is significant in this research because of the greatly increased potential performance for this CPU-bound application. Fortunately, our application is already amenable to parallel solution.

8.7.1 CUDA-Based Solver Introduction

The Compute Unified Device Architecture (CUDA) (NVIDIA 2011) by NVIDIA provides a widely used developer-friendly General Purpose Computing on Graphics Processing Units (GPGPU) interface. CUSP (Bell and Garland 2010) is an open source library for sparse linear algebra computations using CUDA. It provides a flexible, high-level interface for manipulating sparse matrices and solving sparse linear systems. The CUSP library contains two iterative linear solvers, the Conjugate Gradient solver (CG) and Biconjugate Gradient Stabilized solver (BICGS) with an optional Jacobi preconditioner. With a proper construction of the linear system, a simple function call in MATLAB can solve the linear system using the GPU computing power without any prior knowledge about CUDA GPU programming. Indeed, that is one of the messages of this paper: that these tools exist and are useful. Figure 8.5 shows an outline of the solver.

8.7.2 Solver Selection and MATLAB Integration

MATLAB's Cholesky Factorization (CF) exact solver is compute-intensive. However a precise solution is not required since 5D-ODETLAP is lossy. The conjugate gradient (CG) iterative solver in the CUSP library may work better.

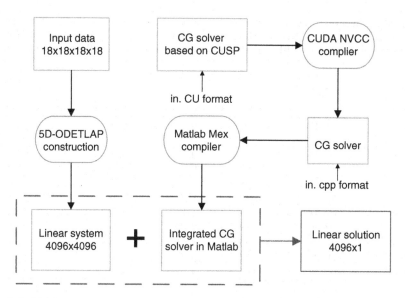

Fig. 8.5 Outline of the solver

First, we tested the CUSP solver on a linear system of size 234256 ×234256 constructed by 5D-ODETLAP. The CUSP CG solver, taking 179 seconds, was much more efficient than the CF direct solver (49,237 s) and CG solver (5,495 s). However, of that 179 seconds, only 9 seconds was spent on actually solving the linear system. The remaining 170 seconds was spent on the data transfer between 5D-ODETLAP, implemented in MATLAB, and the CUSP solver, a C++ executable. To improve this situation, we utilized the MATLAB Executable interface (MEX), which allows users to interface C, C++ or Fortran subroutines to MATLAB. MEX-files are a way to call the custom C, C++ or FORTRAN routines directly from MATLAB as if they were MATLAB built-in functions. Therefore we can largely reduce the overhead for transferring data between the MATLAB program and the CUSP library. But since the CUSP library is implemented both in C++ and CUDA, this adds a certain amount of complexity to incorporate it into the MEX file. Fortunately, CUDA compiler allows users to compile CUDA code into C++ code as an intermediate step using *nvcc –cuda*. So first we write the CUSP code in a MEX style. Then this mixed source code will be compiled into C++ code, which can then be compiled into a MEX file and called directly from the MATLAB program.

In Table 8.2, a comparison of solvers in MATLAB and CUSP demonstrates that for the linear system from 5D-ODETLAP, the CUSP CG solver with Jacobi precon-ditioner runs the fastest. The linear systems in this table are constructed from origi-nal 4D datasets ranging from 8^4 to 18^4, so the resulting linear systems' size ranges from 4096 ×4096 to 104976 ×104976. This solver runs more than seven times faster (39.67:4.84) than its CPU counterpart with the same residual size of their solutions in the test linear system of size 104976 ×104976. Also, it is more than 1340 times

Table 8.2 Effective solver time efficiency comparisons between the MATLAB Cholesky Factorization (CF) direct solver, MATLAB CG solver, CUSP CG solver, CUSP CG solver with Jacobi preconditioner, CUSP BICGS solver with preconditioner and CUSP CG solver. The time measurement is in seconds.

System Size	Cholesky factor. solver	MATLAB CG	CUSP CG	CUSP CG Jacobi	CUSP BICGS Jacobi
4096^2	2.46	0.63	0.23	0.20	0.21
6561^2	6.49	1.11	0.56	0.30	0.36
10000^2	17.43	2.01	0.52	0.48	0.55
14641^2	43.02	3.02	0.75	0.70	0.77
20736^2	93.20	5.55	0.94	0.92	1.08
28561^2	203.25	5.40	1.29	1.28	1.32
38416^2	454.41	13.35	1.91	1.74	1.86
50625^2	816.90	15.58	2.28	2.20	2.34
65536^2	1791.78	29.30	2.89	2.54	3.03
83521^2	3332.05	30.67	3.57	3.57	3.41
104976^2	6488.24	39.67	4.86	4.84	5.14

faster than the CF direct solve in MATLAB. Not only is the CG solver with Jacobi preconditioner from CUSP better in terms of running time, it uses only 13.2 GB main memory and less than 512 MB device memory on GPU on the workstation described above.

8.8 Comparison with JPEG 2000 and 3D-SPIHT

We used the two real world geospatial 5D datasets mentioned above in Sect. 8.1 to test our 5D-ODETLAP framework. We geographically divided these two datasets into 8 different datasets each of size 90 ×180 ×24 ×12 ×4, named WOA05_1, WOA05_2, WOA05_3, WOA05_4, WOA09_1, WOA09_2, WOA09_3 and WOA09_4. This allowed us to test the robustness of 5D-ODETLAP on 8 distinct datasets, to see how sensitive it is to the particular data. The property values were pre-truncated to single precision floats.

Since 3D-SPIHT and JPEG 2000 are among the most popular lossy compression methods, we compared 5D-ODETLAP with them. For a fair comparison, we used binary search in 3D-SPIHT to find a suitable bit rate to produce results with the same mean percentage error in the one test and the same maximum percentage error in the other. The 90 ×180 ×24 ×12 ×4 5D dataset contains 12 ×4 3D datasets of size 90 ×180 ×24. We applied 3D-SPIHT on each of these 3D datasets and measured the error together as a 5D reconstruction in decompression.

Similarly, we used binary search to find a compression ratio in JPEG 2000 to obtain results with the same error in both cases. We also applied JPEG 2000 to each 90 ×180 2D dataset of the overall 24 ×12 ×4 number of 2D datasets. In addition, since JPEG 2000 takes only unsigned 1, 8 or 16 bit integer input, we first used uniform quantization to reduce each input 2D dataset to 16 bit unsigned integer. Thus, all the three methods had the same mean or maximum percentage error on all the 8 test datasets, again for a fair comparison.

Table 8.3, with data from Li (2011), compares the three methods for the same *mean* percentage error. The first observation is that the maximum percentage error for 5D-ODETLAP is much smaller than that of both JPEG 2000 and 3D-SPIHT. This advantage of 5D-ODETLAP is credited to its iterative greedy sampling process, since it eliminates the points with the largest error at each iteration by adding them into the known-point set. The JPEG 2000 and 3D-SPIHT methods do not have this adaptability, and thus produce a much larger maximum percentage error. Second, the compression ratio of 5D-ODETLAP is generally 4.24–9.8 times as large as that of the JPEG 2000 method and 1.13–3.75 times as large as that of the 3D-SPIHT method.

Table 8.4, with data from Li (2011), shows the result of forcing the *maximum* percentage error to be the same for all three methods. The result may be useful in scenarios in which, for example, users need to have a compressed file with guaranteed no more than 10 % error. 5D-ODETLAP's mean error is larger than the others because it spreads out the error more evenly. However 5D-ODETLAP's compression ratio is even larger than in the previous fixed mean percentage error case. Here, the compression ratio of JPEG 2000 and 3D-SPIHT will be considerably worse than 5D-ODETLAP, since the mean error of the compressed file is unnecessarily small. Comparatively, 5D-ODETLAP's iterative sampling process ensures that the points with largest error are always selected, which reduces the maximum error at each step.

8.9   Conclusion and Future Research Plan

5D-ODETLAP demonstrates the efficient application of a massively parallel GPU to an important GIS problem—compressing multidimensional GIS data. 5D-ODETLAP also exploits spatial and temporal redundancies in the data better than previous methods. 5D-ODETLAP's potential impact extends to multidimensional datasets in other domains, such as computational fluid dynamics (CFD). The NVIDIA GeForce 9800GT, the GPU used in this paper, is several years old; we anticipate even better results with a current graphics processor.

Currently 5D-ODETLAP only partially exploits correlations between 4D data layers within a 5D dataset. Also, its coding of the bitmap denoting the points in S, and its compression of the floating values at those points might be improvable. Finally, overdetermined extensions to other PDEs remain to be investigated. Therefore, we expect even better compression of high-dimensional data in the future.

Table 8.3 Compression comparison between 5D-ODETLAP, JPEG 2000 and 3D-SPIHT with the same *mean* percentage error on eight different datasets. The last two columns show the ratio of 5D-ODETLAP's compression ratio over the compression ratio of JPEG 2000 and 3D-SPIHT respectively.

Dataset	% Fixed Mean Err	% Max Err, JPEG 2000	% Max Err, 3D-SPIHT	% Max Err, 5D-ODETLAP	Ratio $\left(\dfrac{5D-ODETLAP}{JPEG2000}\right)$	Ratio $\left(\dfrac{5D-ODETLAP}{3D-SPIHT}\right)$
woa05_1	1.42	44.83	66.46	10.41	8.16	2.40
woa05_2	1.48	49.33	59.18	9.35	9.56	3.61
woa05_3	1.47	65.56	80.23	8.94	4.24	1.13
woa05_4	1.56	67.56	74.14	10.81	8.57	2.41
woa09_1	1.46	48.13	68.18	9.02	8.18	2.41
woa09_2	1.49	51.21	62.15	8.77	9.80	3.75
woa09_3	1.54	75.00	79.35	11.13	4.24	1.14
woa09_4	1.58	65.55	71.50	11.58	8.58	2.42

Table 8.4 Compression comparison between 5D-ODETLAP, JPEG 2000 and 3D-SPIHT with approximately the same *maximum* percentage error on eight different datasets. The last two columns show the ratio of 5D-ODETLAP's compression ratio over the compression ratio of JPEG 2000 and 3D-SPIHT respectively.

Dataset	% Fixed Max Err	% Mean Err, JPEG 2000	% Mean Err, 3D-SPIHT	% Mean Err, 5D-ODETLAP	Ratio $\left(\dfrac{5D-ODETLAP}{JPEG2000}\right)$	Ratio $\left(\dfrac{5D-ODETLAP}{3D-SPIHT}\right)$
woa05_1	10.41	0.52	0.37	1.42	16.33	11.51
woa05_2	9.35	0.32	0.35	1.48	24.02	15.74
woa05_3	8.94	0.26	0.28	1.47	13.84	9.47
woa05_4	10.81	0.37	0.34	1.56	22.23	14.91
woa09_1	9.02	0.39	0.29	1.46	18.97	13.87
woa09_2	8.77	0.31	0.36	1.49	24.43	15.51
woa09_3	11.13	0.36	0.31	1.54	12.05	8.91
woa09_4	11.58	0.39	0.34	1.58	21.92	15.22

Acknowledgements This research was partially supported by NSF grants CMMI-0835762 and IIS-1117277.

References

Anagnostou, K., Atherton, T.J., Waterfall, A.E.: 4d volume rendering with the shear warp factorisation. In: Proceedings of the 2000 IEEE symposium on Volume Visualization, VVS '00, pp. 129–137. ACM, New York, NY, USA (2000). DOI http://doi.acm.org/ 10.1145/353888.353909

Bell, N., Garland, M.: CUSP: Generic Parallel Algorithms for Sparse Matrix and Graph Computations. http://cusp-library.googlecode.com (2010). Version 0.1.0

Bjøke, J.T., Nilsen, S.: Efficient representation of digital terrain models: compression and spatial decorrelation techniques. Computers & Geosciences **28**(4), 433–445 (2002). DOI DOI:10.1016/ S0098-3004(01)00082-6

Franklin, W.R.: The RPI GeoStar project. In: 25th International Cartographic Conference. Paris (2011)

Franklin, W.R., Inanc, M., Xie, Z.: Two novel surface representation techniques. In: Autocarto 2006. Cartography and Geographic Information Society, Vancouver Washington (2006)

Franklin, W.R., Said, A.: Lossy compression of elevation data. In: Seventh International Symposium on Spatial Data Handling. Delft (1996)

Inanc, M.: Compressing terrain elevation datasets. Ph.D. thesis, Rensselaer Polytechnic Institute (2008)

Kidner, D.B., Smith, D.H.: Advances in the data compression of digital elevation models. Computers & Geosciences **29**(8), 985–1002 (2003). DOI DOI:10.1016/S0098-3004(03) 00097-9

Kim, B.J., Pearlman, W.: An embedded wavelet video coder using three-dimensional set partitioning in hierarchical trees (SPIHT). In: Data Compression Conference, 1997. DCC '97. Proceedings, pp. 251–260 (1997). DOI 10.1109/DCC.1997.582048

Lalgudi, H., Bilgin, A., Marcellin, M., Nadar, M.: Compression of fMRI and ultrasound images using 4D SPIHT. In: Image Processing, 2005. ICIP 2005. IEEE International Conference on, vol. 2, pp. II – 746–9 (2005). DOI 10.1109/ICIP.2005.1530163

Li, Y.: CUDA-accelerated HD-ODETLAP: a high dimensional geospatial data compression framework. Ph.D. thesis, Rensselaer Polytechnic Institute (2011)

Li, Y., Lau, T.Y., Stuetzle, C., Fox, P., Franklin, W.R.: 3D oceanographic data compression using 3D-ODETLAP. In: 18th ACM SIGSPATIAL International Conference on Advances in Geographic Information Systems (ACM SIGSPATIAL GIS 2010). San Jose, CA, USA (2010). (PhD Dissertation showcase)

Lloyd, S.: Least squares quantization in PCM. Information Theory, IEEE Transactions on **28**(2), 129–137 (1982). DOI 10.1109/TIT.1982.1056489

Locarnini, R.A., Mishonov, A.V., Antonov, J.I., Boyer, T.P., Garcia, H.E., Baranova, O.K., Zweng, M.M., Johnson, D.R.: World ocean atlas 2009, volume 1: Temperature p. 184 (2010)

Lum, E.B., Ma, K.L., Clyne, J.: Texture hardware assisted rendering of time-varying volume data. In: VIS '01: Proceedings of the conference on Visualization '01, pp. 263–270. IEEE Computer Society, Washington, DC, USA (2001)

Mehlhorn, K., Näher, S.: LEDA: a platform for combinatorial and geometric computing. Commun. ACM **38**(1), 96–102 (1995). http://www.mpi-sb.mpg.de/guide/staff/uhrig/leda.html

Menegaz, G., Thiran, J.P.: Lossy to lossless object-based coding of 3-d mri data. IEEE Transactions on Image Processing **11**(9), 1053–1061 (2002). DOI 10.1109/TIP.2002. 802525

Muckell, J.: Evaluating and compressing hydrology on simplified terrain. Master's thesis, Rensselaer Polytechnic Institute (2008)

NVIDIA: NVIDIA Corporation: Compute Unified Device Architecture Programming Guide. http://developer.nvidia.com/cuda (retrieved 1/11/2011)

Plaza, A., Plaza, J., Paz, A.: Improving the scalability of hyperspectral imaging applications on heterogeneous platforms using adaptive run-time data compression. Computers & Geosciences 36(10), 1283–1291 (2010). DOI DOI:10.1016/j.cageo.2010. 02.009

Sanchez, V., Nasiopoulos, P., Abugharbieh, R.: Lossless Compression of 4D Medical Images using H.264/AVC. In: 2006 IEEE International Conference on Acoustics, Speech and Signal Processing, 2006. ICASSP 2006 Proceedings, vol. 2, p. II (2006). DOI 10.1109/ICASSP.2006.1660543

Stookey, J.: Parallel terrain compression and reconstruction. Master's thesis, Rensselaer Polytechnic Institute (2008)

Stookey, J., Xie, Z., Cutler, B., Franklin, W.R., Tracy, D.M., Andrade, M.V.: Parallel ODETLAP for terrain compression and reconstruction. In: W.G. Aref, et al. (eds.) 16th ACM SIGSPATIAL International Conference on Advances in Geographic Information Systems (ACM GIS 2008). Irvine CA (2008)

Taubman, D.S., Marcellin, M.W., Rabbani, M.: Jpeg2000: Image compression fundamentals, standards and practice. Journal of Electronic Imaging 11, 286 (2002). DOI doi:10.1117/1.1469618

Tracy, D.M.: Path planning and slope representation on compressed terrain. Ph.D. thesis, Rensselaer Polytechnic Institute (2009)

Xie, Z.: Representation, compression and progressive transmission of digital terrain data using over-determined laplacian partial differential equations. Master's thesis, Rensselaer Polytechnic Institute (2008)

Yang, W., Lu, Y., Wu, F., Cai, J., Ngan, K., Li, S.: 4-D wavelet-based multiview video coding. IEEE Transactions on Circuits and Systems for Video Technology 16(11), 1385–1396 (2006)

Ziegler, G., Lensch, H., Magnor, M., Seidel, H.P.: Multi-video compression in texture space using 4d spiht. In: Multimedia Signal Processing, 2004 IEEE 6th Workshop on, pp. 39–42 (2004). DOI 10.1109/MMSP.2004.1436410

Chapter 9
Accelerating Agent-Based Modeling Using Graphics Processing Units

Wenwu Tang

Abstract Agent-based modeling is a disaggregated simulation approach for the exploration of complex spatial dynamics in geographic systems. The use of agent-based models for investigating social-ecological complexity in geographic systems is, however, severely hampered by the computational intensity of agent-based models. Graphics Processing Units (GPUs) are cutting-edge many-core parallel computing platforms that hold great potential in addressing this computational intensity. It is thus necessary to identify aspects that are fundamental in guiding the transformation of agent-based models into GPU environments. The objective of this paper is to identify and discuss the fundamental aspects that need to be considered when using GPUs to accelerate agent-based models. Specifically, these aspects include random number generation, parallelization of agent-based interactions, analysis of agent and environment patterns, and evaluation of computing performance. By linking with these aspects, I used a case study of modeling spatial opinion exchange to illustrate the massively parallel computing power of GPUs for accelerating agent-based modeling. Experimental results suggest that these aspects provide valuable guidance for transforming agent-based models into GPUs to best exploit massively parallel computing power. Further, these aspects are of vital importance for bridging the gap between advancement in GPUs and their applications for resolving spatiotemporal problems using agent-based modeling.

Keywords Agent-based modeling (ABM) • GPU • Parallel computing

W. Tang (✉)
Center for Applied Geographic Information Science, Department of Geography and Earth
Sciences, University of North Carolina at Charlotte, Charlotte, NC 28223, USA
e-mail: WenwuTang@uncc.edu

X. Shi et al. (eds.), *Modern Accelerator Technologies for Geographic Information Science*, 113
DOI 10.1007/978-1-4614-8745-6_9, © Springer Science+Business Media New York 2013

9.1 Introduction

Agent-based modeling is a disaggregated simulation approach for the exploration of complex spatiotemporal dynamics in geographic systems (Epstein and Axtell 1996; Parker et al. 2003; Brown et al. 2005a). Agent-based models (ABMs) have become an appropriate virtual laboratory platform that allows for scenario analysis often intractable for physical experiments. ABMs, with origins from ecology, social science, artificial intelligence, and cognitive science (Epstein and Axtell 1996; Ferber 1999; Grimm and Railsback 2005), have a diverse range of applications in investigating social-ecological complexity. In the GIScience domain, ABMs serve as a flexible modeling approach that has been integrated with GIS (Geographic Information Systems; see Goodchild 1992; Worboys and Duckham 2004) for representing spatially explicit dynamics of geographic systems (Gimblett 2002; Brown et al. 2005b). The integration of GIS and associated spatial analytical capabilities substantially benefits ABMs in terms of using geographically referenced empirical data for model calibration and validation.

The use of ABMs for investigating social-ecological complexity in geographic systems is severely hampered by the computational intensity of ABMs (Wang et al. 2006; Tang and Wang 2009; Wang and Armstrong 2009). A suite of model-level aspects lead to the computationally intensive nature of ABMs. These aspects include stochastic influential factors, nonlinear and concurrent agent-based interactions, iterations for capturing dynamics of systems, and Monte Carlo simulation for inferring modeling outcomes while coping with associated uncertainties (see Wang et al. 2006; Tang and Wang 2009; Tang et al. 2011). The combination of these aspects often makes agent-based modeling data- and computation-intensive. Modelers are increasingly realize the capability of high-performance and parallel computing (Wilkinson and Allen 2004) in resolving the data- and computation-intensities of ABMs. Parallel ABMs have been developed to accelerate the spatially explicit modeling of complex geographic systems (see Abbott et al. 1997; Nagel and Rickert 2001; Wang et al. 2006; Tang and Wang 2009; Tang et al. 2011).

The objective of this paper is to elicit the potential of a massively parallel computing technology, Graphics Processing Units (GPUs; see Owens et al. 2008; Kirk and Hwu 2010), in accelerating ABMs. GPUs represent cutting-edge many-core computing capabilities that have substantially promoted and advanced mainstream computing paradigm (Owens et al. 2007). This study focuses on identifying fundamental aspects that guide the use of GPUs for computationally intensive agent-based modeling. The rest of the paper is organized in the following manner. First, I introduce basics in GPU-accelerated general-purpose computation. Second, I identify and discuss in detail the fundamental aspects that need to be considered when using GPUs to accelerate ABMs. Third, I used a case study to illustrate the massively parallel computing power of GPUs for accelerating agent-based modeling. Fourth, this paper ends with conclusion and discussion on future research themes.

9.2   General-Purpose Graphics Processing Units

GPUs are programmable graphics processors that are built on many-core architecture to provide massively parallel computing power for accelerating general-purpose computation (Owens et al. 2007, 2008; Kirk and Hwu 2010). Since 1980s, GPUs have been developed for the purpose of accelerating graphics operations. APIs (Application Programming Interfaces), such as OpenGL and DirectX, have been available to make GPUs programmable for graphics acceleration (Kirk and Hwu 2010). The potential of GPUs in enabling general-purpose computing has been realized, especially as performance improvement on a single Central Processing Unit (CPU) reached its physical limits. However, the burgeoning of GPUs into general-purpose computing began in 2007 when Nvidia released its parallel computing platform, Compute Unified Device Architecture (CUDA; see CUDA 2013). The use of GPUs for general-purpose computing relies on a multi-threading mechanism for exploiting many-core stream computing power in GPUs (Kirk and Hwu 2010). This makes GPUs capable of high-throughput computing, well-suited to the processing of massive data (data parallelism). A modern CUDA-capable GPU typically comprises multiple streaming multiprocessors each consisting of a collection of streaming processors. For example, the latest Nvidia Tesla K20 GPU, built on the advanced Kepler architecture (Kepler 2013), has in total 2,496 cores in 13 streaming multiprocessors (192 cores per processor). The many-core architecture and multi-threading mechanism lead to the supercomputing power of GPUs at a level of tera-FLOPs (FLoating point Operations Per Second), compared with giga-level floating point performance provided by modern CPUs (Kirk and Hwu 2010). In particular, as computing power in a single GPU grows continually, multi-GPU computing resources that coordinate a cluster of GPU devices become increasingly popular. This creates great opportunities for leveraging GPUs for complex domain-specific problem-solving.

Besides CUDA, a suite of software platforms (e.g., OpenCL and OpenACC; see http://www.khronos.org/opencl/ and http://www.openacc-standard.org/) have been developed for GPU-based programming. These platforms provide interfaces for alternative programming languages (e.g., C/C++, Fortran, and Python). In particular, CUDA is the GPU programming platform that has been most commonly used for general-purpose computation. Standard data structures and algorithms for general-purpose computation are now implemented and available as libraries in CUDA (see https://developer.nvidia.com/gpu-accelerated-libraries). Essentially, CUDA is a scalable programming platform based on a grid-block-thread hierarchy to harness the many-core computing power in GPUs (Kirk and Hwu 2010). Algorithms of a domain-specific model are encapsulated as kernel functions invoking thread-based grids for acceleration. The number of threads allowed in a thread grid can reach 10^{12} (e.g., in the latest Tesla K20 GPU). These threads are grouped into blocks while they are ported into streaming multiprocessors in GPUs. Threads and blocks can be organized into different dimensions (1D, 2D, and 3D) according to data characteristics and algorithmic design. Alternative levels of memory (e.g., local, shared, and global) are available for maintaining data at different thread levels (thread, block, and grid). The hierarchical multi-threading design in CUDA allows for using multi-level parallelism to achieve the high scalability of domain-specific models transformed onto GPUs.

9.3 Accelerating Agent-Based Models Using Graphics Processing Units

ABMs rely on the concept of agents to represent real-world entities (e.g., decision makers) that interact with their peers and environments (Epstein 1999; Ferber 1999; Grimm and Railsback 2005). Agents employ rules to characterize their interacting behavior in response to change in system dynamics. Regarding model development, agents are implemented as objects interacting with spatially explicit environments that are continuous or discretized (i.e., vector- or raster-based environmental representation; see Worboys and Duckham 2004; Brown et al. 2005b). Agent-based interactions (including agent-agent and agent-environment) are often decentralized, nonlinear, and self-organized. These interactions drive the emergence of high-level patterns in geographic systems. Model configuration (e.g., the number of agents, spatial extent, and temporal duration) of most CPU-based sequential ABMs is limited because of computational burden induced by calibration and validation. For example, the spatial extent of most raster-based ABMs is often smaller than $1,000 \times 1,000$ cells. This greatly inhibits the use of ABMs for investigating geographic systems often driven by multi-scale influential factors.

Because agent-based interactions are decentralized and concurrent, ABMs have great potential to leverage many-core parallel computing capabilities in GPUs. Though general-purpose GPU technologies emerged several years ago, a suite of modeling efforts have been reported to investigate the use of GPUs for accelerating ABMs. Table 9.1 summarizes existing modeling efforts in terms of agent types (plants, animals, and human decision makers). These GPU-enabled ABMs were developed to address research questions (e.g., collective behavior or movement) from different domains, including computer science, ecology, social science, and transportation. Note that in this study, cellular automata are regarded as a special form of ABMs in which agents are immobile and interact with their spatial neighborhood (see Ferber 1999). CUDA is the software platform mostly chosen to implement ABMs for GPU acceleration. Programming languages mainly include C/C++ and Java. The massively parallel computing power in GPUs allows modelers to further extend model configuration. For example, the number of agents allowed reaches a level of millions or billions. Environmental representation in these existing GPU-enabled ABMs is vector- (including networks) or raster-based.

The key challenge for using GPUs to accelerate ABMs is how to transform an ABM into GPU-enabled environments (see Fig. 9.1 for illustration). Most of the sequential ABMs were developed specifically to cope with the CPU-based

Table 9.1 Summary of agent-based models accelerated using GPUs

Category	Citation	Modeling questions
Plant species	van de Koppel et al. (2011)	Pattern formation of tiger bush
	Keenan et al. (2012)	Tree dispersal and competition
Animal species	Passos et al. (2009)	Bird flocking
	Li et al. (2009)	Fish schooling
	Erra et al. (2009)	
	van de Koppel et al. (2011)	Mussel disturbance
Human decision makers	Perumalla et al. (2009)	Mood diffusion
		Residential segregation
	Richmond and Romano (2008)	Pedestrian movement
	Chen et al. (2011)	Crowd gathering
	Tang and Bennett (2011)	Opinion exchange
	Strippgen and Nagel (2009)	Vehicle mobility
	Wang and Shen (2012)	

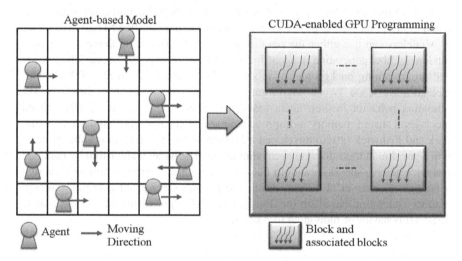

Fig. 9.1 Transformation of an agent-based model into GPUs (an example of mobile agents was used)

computing environments. These CPU-enabled ABMs cannot be directly run within the GPU environments. The transformation of ABMs into GPUs requires the consideration of fundamental aspects that couple modeling and computing domains. These aspects include (but are not limited to): random number generation for stochastic modeling, the mapping of agent-based interactions to GPU-supported threads, the analysis of agent and environment patterns, and the evaluation of acceleration performance. In this section, I discuss in detail these aspects.

9.3.1 Random Number Generation for Stochastic Modeling

In ABMs of geographic systems, agent-based interactions are driven by influential factors that are often uncertain—i.e., these ABMs are stochastic. Random number generation is thus required for stochastic agent-based modeling. Random number generation libraries have been well developed within CPU environments. However, these CPU-enabled libraries of sequential random number generation cannot be directly used within GPU environments. Random number generation algorithms have to be re-implemented for GPU-enabled ABMs. Two approaches exist for generating random numbers in GPUs (see Li et al. 2009). One approach is to generate random numbers in CPU memory first, and then transfer these numbers to GPU device memory. This approach is relatively straightforward since it is based on capabilities already available within CPU environments. The other approach is to directly generate random numbers on GPUs. While CUDA now provides support for generating random numbers in the CURAND library (CURAND 2013), early versions of CUDA toolkits do not have such capabilities. The CURAND library supports multiple random number generation algorithms, including xor-shift, combined multiple recursive, and Mersenne Twister families (CURAND 2013).

Modelers can use either the parallel library in CUDA or implement their own random number generation in agent-based modeling, contingent on modeling needs and programming background of modelers. For example, Li et al. (2009) implemented a Mersenne Twister algorithm to generate random numbers that drive fish schooling behavior in their agent-based fish simulation. Li et al. (2009) utilized block-level shared memory to improve the parallel generation of random numbers. Park and Fishwick (2010) developed a Mersenne Twister random number generation algorithm in their discrete event simulation. The parallel algorithm by Park and Fishwick (2010) supports random numbers following alternative probabilistic distributions. Keenan et al. (2012) developed parallel GPU-enabled pseudo-random number generation based on the Tausworthe algorithm to simulate the stochastic dispersal behavior of individual tree species.

9.3.2 Parallelize Agent-Based Interactions
Using Thread Parallelism

Interactions in ABMs comprise agent-agent, agent-environment, and environment-environment interactions. Update of these interactions leads to the computational intensity of ABMs. The key to the use of GPUs for accelerating ABMs is to map these interactions and data structures that represent agents and environments to thread and memory hierarchies on GPUs. Because of massive concurrent threads supported, GPUs have great potential to empower the computation of massive updates induced by interactions in ABMs. This makes it possible to conduct GPU-enabled agent-based modeling in which the number of agents, the length of temporal iterations, or the number of environmental features reach several order of magnitude of those developed within CPU environments. How to map interactions in ABMs to threads in GPUs (i.e., domain decomposition in parallel computing; see Ding and Densham 1996; Wilkinson and Allen 2004) is dependent on, for example, the number of agents used and the computational intensity of agent-level interactions. In a large ABM enabled by GPUs, each thread can handle interactions associated with a small sub-set of agents, given that agent-based interactions are light-weighted. For ABMs in which agent-based interactions are highly intensive (the number of agents may be limited), updates on an agent's interactions can be mapped to a single thread or multiple threads.

GPUs are based on multi-threading and shared-memory mechanisms for parallel computing. When the set of threads updates interactions in an ABM, the state of an agent or environmental feature is likely to be accessed and modified by multiple threads. CUDA supports simultaneous access to the state of an agent. However, an issue of mutual exclusion (Raynal 1986; Anderson et al. 2003) arises when multiple threads attempt to modify (write operation) the state of an agent or environmental feature simultaneously. This causes the incorrect update of agent or environment state. To resolve this issue, atomic functions, which ensure an agent or environment state cannot be modified by another thread while it is updated by a thread, are often needed. For example, Keenan et al. (2012) used an atomic compare-and-swap function to support the establishment of new trees in their individual-based model of tree dispersal and competition.

The iterative nature of ABMs requires that all interactions within a discrete time step are completed before entering next step—i.e., synchronization at each time step is required. While we use many concurrent threads to update interactions in ABMs, we need to make sure the update of these interactions is synchronized at each time step. CUDA does not support the synchronization of grid-level threads (the entire collection of threads recruited) within a kernel function. However, when different kernel functions are invoked or a kernel function is invoked multiple times, the synchronization of grid-level threads is achieved. Thus, iterations of ABMs in GPUs can be implemented by repetitively invoking kernel functions corresponding to model iterations. Further, threads within a block can be synchronized in a kernel function in CUDA. Thus, modelers can synchronize agent-based updates handled by threads within a block. The two levels of synchronization (grid- and block-level) have been commonly used in most existing ABMs accelerated using GPUs.

9.3.3 Analysis of Agent or Environmental Patterns

ABMs are often associated with a significant amount of agent- or environment-related patterns (simulated or observed). Quantitative metrics (e.g., spatial or non-spatial, local or global, and spatial or spatiotemporal) can be used to evaluate these patterns (see Brown et al. 2005a). To derive the metrics of agent- or environment-related patterns often requires the traversal of the entire set of agents or environmental features. Depending on the size of agent population or environmental features and the computational intensity of metrics, the computation of these metrics can be allocated on CPUs (small size and low computational intensity) or GPUs (large size or high computational intensity).

Parallel scan or reduction algorithms (see Harris et al. 2007) exist to support the concurrent and efficient computation of these metrics. For example, in their Schelling segregation model, Perumalla and Aaby (2008) used parallel reduction for counting the number of agents to detect whether a residential system reaches equilibrium. Richmond and Romano (2008) conducted summary statistics in their agent-based pedestrian model with support from parallel reduction operations. Chen et al. (2011) applied the metric of information entropy into their large crowd simulation to evaluate the diversity of behavioral types of pedestrians. Keenan et al. (2012) developed a GPU-enabled parallel approach to derive the crowding index of trees in their individual-based model using parallel reduction operations available in the CUDA thrust library.

9.3.4 Evaluation of Computing Performance

The evaluation of computing performance of GPU-accelerated ABMs lies in the comparison of computing time between GPU-enabled parallel ABMs with CPU-based sequential counterparts. A commonly used performance metric is acceleration factor (Preis et al. 2009), which is the ratio of CPU execution time over GPU time. Although it is conceptually similar to the metric of speed up in CPU-based parallel computing (Wilkinson and Allen 2004), acceleration factor has reliance on both GPU and CPU configuration. The metric of acceleration factor allows for examining variation in computing performance in response to change in model parameters or CUDA configuration. Acceleration factors of most ABMs reported in the literature reach at least one order of magnitude, depending on the complexity of agent-based interactions and the problem size with which ABMs are associated (e.g., the size of agent populations).

Li et al. (2009) simulated the schooling behavior of fishes, driven by attraction or repulsion influence from neighboring fishes. The acceleration factor obtained by Li et al. (2009) is around 230–240, given 100 fishes simulated on NVIDIA GeForce 8800 GTX (128 cores; see http://www.nvidia.com/page/geforce_8800.html) for GPU against Intel Pentium CPU. Passos et al. (2009) examined the computing performance of their agent-based flocking boid model. Passos et al. (2009) compared GPU and CPU computing time over the number of flocking boids. A nonlinear response of computing performance was observed when the number of boids changes from 64 to 1 million. Correspondingly, acceleration factors (calculated from results reported by Passos et al. (2009)) increase from 0.283 to 138.464. Acceleration factors tend to increase rapidly when the number of agents increases from 64 to 262,144. Besides the number of agents, rules that represent agent-based interactions may have substantial influence on the computing performance of GPU-accelerated ABMs. In Keenan et al. (2012), acceleration factors increase from about 60 to 360 when the crowding effect of tree species was computed. These reported performance results suggest that computationally intensive agent-based updates in ABMs can really reap benefits from GPUs.

9.4   Case Study

In this section, I used a parallel agent-based model of spatial opinion exchange to investigate the potential of GPUs for accelerating spatial agent-based modeling. This GPU-enabled model has been detailed in Tang and Bennett (2011). In this study, I focus on using this model to demonstrate the strength of GPUs for accelerating ABMs by linking with the fundamental aspects highlighted in Sect. 9.3.

9.4.1   Parallel Agent-Based Opinion Modeling

The agent-based opinion model was designed to investigate how spatially aware decision makers communicate among themselves to develop consensus on topics of interest. The consensus development represents a complex space-time diffusion process in which the communication behavior of individuals and their spatial distributions play a pivotal role. In this model, individuals are simulated as agents situated on a raster-based 2D landscape. An agent exchanges opinion (as in a continuous normalized variable; i.e., in the range of [0, 1]) with her/his neighbors, and this opinion exchanging behavior is characterized by distance-decayed spatial neighborhood search and opinion update (driven by a bounded confidence model; see Weisbuch et al. 2002). Each iteration, an agent will randomly pick a neighboring agent within her/his influential range to decide whether the two agents will update opinions based on an opinion threshold (in the range of [0, 0.5]; large values represent open-mindedness). If opinion distance between the two agents is less than the opinion threshold, the opinion of the current agent will be modified using the bounded confidence model.

The agent-based opinion model has been parallelized using GPUs to cope with large agent populations. Random numbers in this model were produced at thread level based on a linear congruential generator (Gentle 2003). Seeds of these thread-level random number sequences were generated and fed from CPU. A probability, denoted as p, for a thread is obtained using the following equations.

$$p = RND_n \ / \ rmax \tag{9.1}$$

$$RND_n = \left(a^* RND_{n-1} + b\right) \bmod rmax \tag{9.2}$$

where RND_n and RND_{n-1} are random numbers generated at the nth and $n-1$th iteration. RND_0 is assigned by a random number transferred from CPU. $rmax$ is the maximal random number allowed ($rmax = 232$ in this study). a and b are coefficients to generate the nth random number based on the $n-1$th random number ($a = 1,664,525$ and $b = 1,013,904,223$ in this study). "mod" without quotation refers to the modulus operator.

In the parallel model, a kernel function was developed to recruit threads for opinion exchange among agents. To update interactions for opinion exchange needs the traversal of the entire agent population each iteration. Thus, domain decomposition is based on the partitioning of the entire agent population and assigning partitioned sub-populations to CUDA threads. Because the number of threads that CUDA supports is large (giga-level), each thread only needs to handle a small group of agents (one-to-many relationship between threads and agents) even for the model with large agent population. The spatial neighborhood search and opinion update processes of agents are decentralized and concurrent. It is likely that an agent's opinion may be updated by two or more agents from different threads simultaneously. Consequently, the opinion of the agent may not be updated correctly—a mutual-exclusion issue. Atomic operations were used to resolve this mutual-exclusion issue. Further, to assure the opinions of all agents are updated each iteration synchronously, the kernel function of opinion exchange is invoked each iteration.

Two metrics, Shannon's entropy and Simpson index, were used to evaluate the degree of opinion diversity of agents. These metrics are based on the statistical distribution of agent opinions (Hill 1973; Keylock 2005). Equations 9.3 and 9.4 show the derivation of the Shannon's entropy and Simpson index.

$$entropy = -\sum_{i=1}^{n} p_i^* \ln(p_i) \tag{9.3}$$

$$simpson = \sum_{i=1}^{n} p_i^2 \tag{9.4}$$

where $entropy$ and $simpson$ are the Shannon's entropy and Simpson index of agent opinion. p_i denotes the proportion of agents in opinion group i. n represents the number of opinion groups. Low entropy or high Simpson index represents a homogeneous opinion pattern (i.e., the opinion system reaches consensus). While these two metrics allow for evaluating modeling performance, acceleration factor was used to examine the computing performance of the model.

9.4.2 Experiment and Discussion

I designed an experiment to examine the impact of the proportion of open- and narrow-minded agents on the development of consensus. Two types of agents, open- and narrow-minded, were used. Open-minded agents have a large opinion threshold

Fig. 9.2 Spatial patterns of agent opinion and types (landscape size: 3,000×3,000; ratio of open-minded over narrow-minded agents: 1:1; agent opinion was scaled to the range of 0–1,000)

(0.26) and a long communication range for spatial neighborhood search. Narrow-minded agents are characterized by a relatively small opinion threshold (0.20) and a short communication range. All agents' opinions were initialized between 0 and 1. I varied systematically the proportion of open-minded agents from 95 % down to 5 % at an interval of 5 %, in total 19 treatments (noted as T1–T19). Landscape size is 3,000×3,000. Each cell is situated by one agent. The type (open- or narrow-minded) of the agent in a cell was determined randomly, and each agent has equal probability to be either open- or narrow-minded. Figure 9.2 is an example showing spatial patterns of agent type and initial opinion. The number of iterations is 1,000. The number of model repetitions for each treatment is 20. Regarding GPU configuration, the number of threads per block is 256. Thus, the number of thread blocks is 35,157. The GPU device is Tesla M2050 (448 cores; 1.15 GHz of clock rate; 3 GB global memory). Intel Xeon processors with 2.67 GHz of clock rate and 12 GB of memory were CPUs for sequential execution.

Table 9.2 reports results of computing performance of the 19 treatments. Computing time and acceleration factors are based on averaged results of 20 model repetitions for each treatment. The GPU execution time for a model run (memory transferring time is included) is about 5–6 min. However, the sequential CPU time of a model run is about 1 h. Acceleration factors for the 19 treatments vary around 10–11. Further, execution time for both GPU- and CPU-enabled models tends to increase as the proportion of open-minded agents decreases.

Figure 9.3 shows the spatial patterns of agent opinion at iteration 1 and 1,000 for treatment T1 (95 % open-minded agents) and T19 (5 % open-minded agents). The opinion system dominated by open-minded agents reaches consensus, but for the system with low percentage of open-minded agents, agent opinions remain diverse after 1,000 iterations (i.e., consensus was not obtained). Figure 9.4 depicts results of Shannon's entropy and Simpson index over iterations for the 19 treatments.

Table 9.2 Results of computing performance of the experiment (time units: seconds; T1–T19: treatments for the proportion of open-minded agents varying from 95 to 5 % at an interval of 5 %)

Treatment	GPU time	CPU time	Acceleration factor
T1	338.39	3,507.14	10.36
T2	339.85	3,556.21	10.46
T3	336.90	3,490.12	10.36
T4	336.73	3,598.97	10.69
T5	335.54	3,543.13	10.56
T6	334.35	3,559.00	10.64
T7	335.10	3,705.77	11.06
T8	337.38	3,665.08	10.86
T9	341.11	3,729.63	10.93
T10	338.61	3,814.69	11.27
T11	341.30	3,716.15	10.89
T12	343.33	3,804.65	11.08
T13	343.47	3,882.67	11.30
T14	348.78	3,896.29	11.17
T15	348.93	3,915.16	11.22
T16	348.72	3,965.70	11.37
T17	357.95	3,998.21	11.17
T18	352.17	3,876.38	11.01
T19	358.08	3,903.55	10.90

In general, entropy exhibits a decreasing trend over iteration, but Simpson index shows an increasing pattern. This is because agents interact with each other to modify their opinions based on influence from others. Initially, agent opinions are randomly distributed, corresponding to a most diverse opinion pattern (entropy is high and Simpson index is low). Through the opinion update process, agents tend to move close on their opinion space. Consequently, a decreasing (increasing) pattern of entropy (Simpson index) was observed. A more interesting pattern is that agent opinions in a population with a large number of open-minded agents (e.g., treatment T1 with 95 % open-minded agents) tend to converge to consensus quickly. As the number of open-minded agents' declines, agent opinions tend to converge slowly. As an extreme case, when the proportion of open-minded agents is 5 %, no convergence reaches (also see Fig. 9.3b, d).

Figure 9.5 depicts the number of iterations at which agent opinions converge for the 19 treatments. The convergence speed of agent opinions demonstrates a nonlinear increasing pattern with a threshold at about 40 % in terms of the proportion of open-minded agents (treatment T12). When the proportion of open-minded agents is higher than 40 %, the number of iterations at which agent opinions converge increases slowly from 130 to 290. However, once this proportion is lower than 40 %, the opinion system tends to converge more slowly and even no convergence reaches. This indicates that the proportion of open-minded agents, or generally agent types, in agent population is important in driving consensus development.

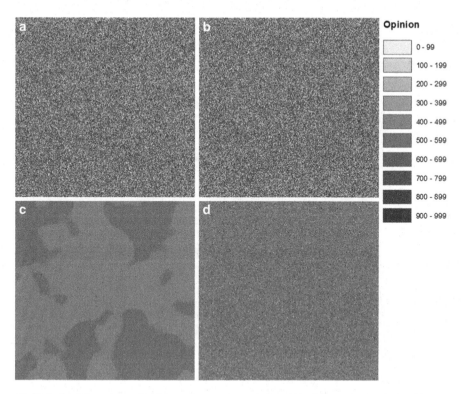

Fig. 9.3 Spatial patterns of opinions for agent population with different proportions of open-minded agents (**a** and **b**: maps of initial opinions for agent population with 95 and 5 % open-minded agents; **c** and **d**: maps of final opinions for agent population with 95 and 5 % open-minded agents; number of iterations: 1,000)

Results of computing performance indicate that GPUs can accelerate considerably the agent-based opinion model. Because the high proportion of open-minded agents leads to the quick convergence of the opinion system, the amount of agent-based interactions associated with computational workload is low. Thus, the computing time for both GPU- and CPU-enabled opinion models tends to be short when the proportion of open-minded agents is high. This demonstrates the importance of appropriately mapping and parallelizing agent-based interactions within GPU environments.

9.5 Conclusion

This study presented the use of GPUs for accelerating ABMs of complex geographic systems. The massively parallel computing power of GPUs for general-purpose computation has been admitted as GPUs' hardware performance almost

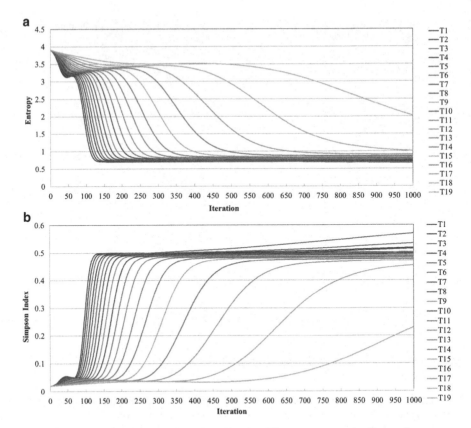

Fig. 9.4 Opinion diversity indices over iterations for different treatments (**a**: Shannon's entropy; **b**: Simpson index; T1–T19: treatments for the proportion of open-minded agents varying from 95 to 5 % at an interval of 5 %)

doubles every year (Kirk and Hwu 2010). GPU resources tend to be ubiquitously available on alternative computing platforms, including mobile devices, desktop computers, clusters, and supercomputers. This trend has greatly motivated modelers to harness parallel computing power in GPUs for domain-specific modeling.

The use of GPUs for parallel agent-based modeling often requires a solid understanding of ABMs, software and hardware characteristics of GPUs. Researchers with collective knowledge from both modeling and computational domains will greatly benefit from the supercomputing capabilities of GPUs. This study identified and discussed fundamental aspects in the exploitation of GPUs for accelerating ABMs. These aspects include random number generation, parallelization of agent-based interactions, analysis of agent and environment patterns, and evaluation of computing performance. The consideration of these aspects will facilitate the transformation of sequential ABMs into GPU-enabled parallel counterparts.

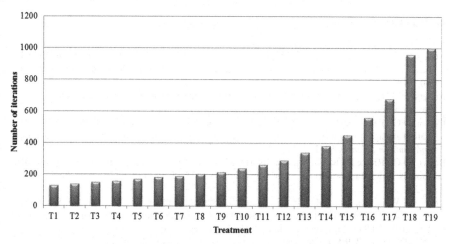

Fig. 9.5 Number of iterations at which agent opinions converge (T1–T19: treatments for the proportion of open-minded agents varying from 95 to 5 % at an interval of 5 %)

However, the application of GPUs for accelerating ABMs still remains in its early stage, suggesting a thread of future research themes. First, there is a set of spatiotemporal problems (e.g., global- or regional-level land cover change, landscape-level mosquito dynamics) often requiring large agent-based modeling. It is imperative to identify concurrent characteristics in these large models and then transform them into many-core GPU-enabled computing environments to gain high acceleration. Second, it is necessary to develop and examine scalable parallel spatial strategies (e.g., spatial domain decomposition and synchronization) for agent-based modeling to efficaciously exploit massively parallel computing power in GPUs. Third, heterogeneous multi-GPU resources (i.e., GPU clusters) are increasingly available while single GPUs' performance keeps improving. To better exploit these multi-GPU resources for large agent-based modeling hinges on the combination of CPU-based parallel programming architectures or platforms (e.g., MPI, OpenMP) with GPU programming environments (e.g., CUDA and OpenCL). While this exploitation may require more knowledge from heterogeneous parallel computing, the size and complexity of the spatiotemporal problems represented by ABMs are enhanced significantly.

Acknowledgements The author thanks support from US NSF Human Social Dynamics #0624292—Collaborative Research: AOC Social Complexity and the Management of the Commons and Faculty Research Grant at the University of North Carolina at Charlotte. University Research Computing (URC) at the University of North Carolina at Charlotte provided partial computing resources for this study. The author would like to thank Meijuan Jia, Jing Deng, Wenpeng Feng, Huifang Zuo, and Dr. Jian Gong for their assistance in manuscript preparation.

References

Abbott, C.A., Berry, M.W., Comiskey, E.J., Gross, L.J. and Luh, H., 1997, Parallel individual-based modeling of Everglades deer ecology. *Computational Science & Engineering, IEEE*, 4, 60–78.

Anderson, J.H., Kim, Y.-J. and Herman, T., 2003, Shared memory mutual exclusion: major research trends since 1986. *Distributed Computing*, 16, 75–110.

Brown, D.G., Page, S., Riolo, R., Zellner, M. and Rand, W., 2005a, Path dependence and the validation of agent-based spatial models of land use. *International Journal of Geographical Information Science*, 19, 153–174.

Brown, D.G., Riolo, R., Robinson, D.T., North, M. and Rand, W., 2005b, Spatial process and data models: Toward integration of agent-based models and GIS. *Journal of Geographic Systems*, 7, 1–23.

Chen, D., Wang, L., Tian, M., Tian, J., Wang, S., Bian, C. and Li, X., 2011, Massively parallel modelling & simulation of large crowd with GPGPU. *The Journal of Supercomputing*, 1–16.

CUDA, 2013, CUDA. http://www.nvidia.com/object/cuda_home_new.html.

CURAND, 2013, CURAND. http://docs.nvidia.com/cuda/curand/index.html.

Ding, Y.M. and Densham, P.J., 1996, Spatial strategies for parallel spatial modelling. *International Journal of Geographical Information Systems*, 10, 669-698.

Epstein, J.M., 1999, Agent-based computational models and generative social science. *Complexity* 4, 41–60.

Epstein, J.M. and Axtell, I., 1996, *Growing Artificial Societies: Social Science from the Bottom Up* (Cambridge: The MIT Press).

Erra, U., Frola, B., Scarano, V. and Couzin, I., 2009, An efficient GPU implementation for large scale individual-based simulation of collective behavior. In *High Performance Computaitonal Systems Biology (HiBi09), October 14–16, 2009* (Trento, Italy).

Ferber, J., 1999, *Multi-agent Systems: An Introduction to Distributed Artificial Intelligence* (New York: Addison-Wesley).

Gentle, J.E., 2003, *Random number generation and Monte Carlo methods* (Springer).

Gimblett, R.H., 2002, Integrating geographic information systems and agent-based technologies for modeling and simulating social and ecological phenomena. In *Integrating Geographic Information Systems and Agent-based Modeling Techniques for Simulating Social and Ecological Processes*, R.H. Gimblett (Ed.), 1-20 (Oxford: Oxford University Press).

Goodchild, M.F., 1992, Geographical Information Science. *International Journal of Geographical Information Systems*, 6, 31–45.

Grimm, V. and Railsback, S.F., 2005, *Individual-based Modeling and Ecology* (Princeton, NJ: Princeton University Press).

Harris, M., Sengupta, S. and Owens, J.D., 2007, Parallel prefix sum (scan) with CUDA. *GPU Gems*, 3, 851–876.

Hill, M.O., 1973, Diversity and evenness: a unifying notation and its consequences. *Ecology*, 54, 427–432.

Joselli, M., Passos, E. B., Zamith, M., Clua, E., Montenegro, A., and Feijó, B., 2009, *A Neighborhood Grid Data Structure for Massive 3D Crowd Simulation on GPU*. In 2009 VIII Brazilian Symposium on Games and Digital Entertainment (SBGAMES) (pp. 121–31). Brazil: IEEE.

Keenan, M., Komarov, I., D'Souza, R.M. and Riolo, R., 2012, Novel graphics processing unit-based parallel algorithms for understanding species diversity in forests. In *Proceedings of the 2012 Symposium on High Performance Computing*Society for Computer Simulation International), 10.

Kepler, 2013, Nvidia Kepler Architecture. http://www.nvidia.com/object/nvidia-kepler.html.

Keylock, C., 2005, Simpson diversity and the Shannon–Wiener index as special cases of a generalized entropy. *Oikos*, 109, 203–207.

Kirk, D.B. and Hwu, W.-m., 2010, *Programming Massively Parallel Processors: A hands-on Approach* (Burlington, MA, USA: Morgan Kaufmann).

Li, H., Kolpas, A., Petzold, L. and Moehlis, J., 2009, Parallel simulation for a fish schooling model on a general-purpose graphics processing unit. *Concurrency and Computation: Practice and Experience*, 21, 725–737.

Nagel, K. and Rickert, M., 2001, Parallel implementation of the TRANSIMS micro-simulation. *Parallel Computing*, 27, 1611–1639.

Owens, J.D., Houston, M., Luebke, D., Green, S., Stone, J.E. and Phillips, J.C., 2008, GPU Computing. *Proceedings of the IEEE*, 96, 879–899.

Owens, J.D., Luebke, D., Govindaraju, N., Harris, M., Krüger, J., Lefohn, A.E. and Purcell, T.J., 2007, A Survey of General-Purpose Computation on Graphics Hardware. *Computer Graphics Forum*, 26, 80–113.

Park, H. and Fishwick, P.A., 2010, A GPU-Based Application Framework Supporting Fast Discrete-Event Simulation. *SIMULATION*, 86, 613–628.

Parker, D.C., Manson, S.M., Janssen, M.A., Hoffmann, M.J. and Deadman, P., 2003, Multi-agent systems for the simulation of land-use and land-cover change: A review. *Annals of the Association of American Geographers*, 93, 314–337.

Passos, E.B., Joselli, M., Zamith, M., Clua, E.W.G., Montenegro, A., Conci, A. and Feijo, B., 2009, A bidimensional data structure and spatial optimization for supermassive crowd simulation on GPU. *Computers in Entertainment (CIE)*, 7, 60.

Perumalla KS and Aaby BG (2008) Data parallel execution challenges and runtime performance of agent simulations on GPUs. In Proceedings of the 2008 Spring Simulation Multiconference. (ed.), Vol. pp. 116–23, International Society for Computer Simulation, Ottawa, Canada.

Perumalla, K.S., Aaby, B.G., Yoginath, S.B. and Seal, S.K., 2009, GPU-based real-time execution of vehicular mobility models in large-scale road network scenarios. In *Proceedings of the 2009 ACM/IEEE/SCS 23rd Workshop on Principles of Advanced and Distributed Simulation* IEEE Computer Society), 95–103.

Preis, T., Virnau, P., Paul, W. and Schneider, J.J., 2009, GPU accelerated Monte Carlo simulation of the 2D and 3D Ising model. *Journal of Computational Physics*, 228, 4468–4477.

Raynal, M., 1986, *Algorithms for Mutual Exclusion* (Cambridge, Massachusetts: The MIT Press).

Richmond, P. and Romano, D.M., 2008, A high performance framework for agent based pedestrian dynamics on gpu hardware. In *EUROSIS ESM 2008 (European Simulation and Modelling)* (Le Havre, France.

Strippgen, D. and Nagel, K., 2009, Multi-agent traffic simulation with CUDA. In *High Performance Computing & Simulation, 2009. HPCS'09. International Conference on*IEEE), 106–114.

Tang, W. and Bennett, D.A., 2011, Parallel agent-based modeling of spatial opinion diffusion accelerated using graphics processing units. *Ecological Modelling*, 222, 3605–3615.

Tang, W., Bennett, D.A. and Wang, S., 2011, A parallel agent-based model of land use opinions. *Journal of Land Use Science*, 6, 121–135.

Tang, W. and Wang, S., 2009, HPABM: A Hierarchical Parallel simulation framework for spatially-explicit Agent-Based Models. *Transactions in GIS*, 13, 315–333.

van de Koppel, J., Gupta, R. and Vuik, C., 2011, Scaling-up spatially-explicit ecological models using graphics processors. *Ecological Modelling*, 222, 3011–3019.

Wang, D., Berry, M.W., Carr, E.A. and Gross, L.J., 2006, A parallel fish landscape model for ecosystem modeling. *Simulation*, 82, 451–465.

Wang, K. and Shen, Z., 2012, A GPU based traffic parallel simulation module of artificial transportation systems. In *Service Operations and Logistics, and Informatics (SOLI), 2012 IEEE International Conference on*, 160–165.

Wang, S. and Armstrong, M.P., 2009, A theoretical approach to the use of cyber infrastructure in geographical analysis. *International Journal of Geographical Information Science*, 23, 169–193.

Weisbuch, G., Deffuant, G., Amblard, F. and Nadal, J.-P., 2002, Meet, discuss, and segregate! *Complexity*, 7, 55–63.

Wilkinson, B. and Allen, M., 2004, *Parallel Programming: Techniques and Applications Using Networked Workstations and Parallel Computers (Second Edition)* (Upper Saddle River, NJ USA: Pearson Prentice Hall).

Worboys, M. and Duckham, M., 2004, *GIS: A Computing Perspective, Second Edition* (Boca Raton: CRC Press).

Part IV
MAT in Remotely Sensed Data Processing and Analysis

Chapter 10
Large-Scale Pulse Compression for Costas Signal with GPGPU

Bin Zhou, Chun-mao Yeh, Wen-wen Li, and Wei-jie Zhang

Abstract Costas signal, which is able to provide good time and frequency resolution simultaneously, is widely adopted in modern radar systems, especially for the radar with low interception performance. However, because of the widely acknowledged superb ambiguity properties, the pulse compression for Costas signal requires multi-velocity-channel processing, and hence, the computational load is increased accordingly. In this paper, based on the characteristics of "General Purpose Graphics Processing Unit (GPGPU)", a new pulse compression scheme for Costas signal is proposed and validated. According to the experiment, this processing scheme will provide a good reference for the design of radar signal processors with high computing performance.

Keywords High speed radar signal processing • Costas signals • Multi-velocity channel processing • General purpose graphics processing unit • GPGPU

B. Zhou (✉)
Institute of Oceanographic Instrumentation, Shandong Academy of Sciences, Qingdao, China
e-mail: synosy@gmail.com

C.-m. Yeh
Beijing Institute of Radio Measurement, Beijing, China
e-mail: chunmaoyeh@gmail.com

W.-w. Li
GeoDa Center for Geospatial Analysis and Computation,
Arizona State University, Tempe, AZ, USA
e-mail: wenwen@asu.edu

W.-j. Zhang
Science and Technology on Electronic Information Control Laboratory, Chengdu, China
e-mail: arrowbit@gmail.com

X. Shi et al. (eds.), *Modern Accelerator Technologies for Geographic Information Science*,   133
DOI 10.1007/978-1-4614-8745-6_10, © Springer Science+Business Media New York 2013

10.1 Introduction

Pulse compression is a widely-used technique in modern radar, sonar and communication systems. It can obtain high distance resolution, while maintaining acceptable signal to noise ratio. Pulse compression technique modulates the transmitting pulse and then calculates the correlation between received and modulated pulses. The common waveforms are chirp signal, nonlinear frequency modulation and phase encoding pulse. Linear frequency modulation pulse has distance-speed coupling effect, which leads to serious measuring errors. But in phase encoding, Barker code with the length of more than 13 has not yet been found and pulse compression ratio does not exceed 22.3 dB. In addition, although pseudorandom sequences can obtain higher pulse compression ratio, its range side-lobe is also higher. This affects the following signal detection, such as described in paper (Wenhua and Yixian 1996).

Phase encoding technique has several advantages: firstly, it can obtain high pulse compression ratio, which is needed in any outstanding pulse compression system.

Secondly, because of obvious "thumbtack" characteristic in its ambiguity graph, it can achieve higher velocity resolution and still keeps good range resolution. Finally, the signal possesses better anti-jamming ability (Wenhua and Yixian 1996). Costas signal also draws attention in acoustic applications because of these good properties (Zhiguang et al. 2010). However, when using Costas signal on many speed channels, the time and space complexity are huge. Single processor cannot handle it. Hence it requires Massively Parallel Processing (MPP) (Hongxun and Xiuchun 2004). The ordinary MPP system includes distributed share memory (DSM), parallel processor array (PPA), cluster system and so on. Cluster system is low-cost, flexible and expandable. But it has large size and high power consumption. PPA and DSM systems are smaller and low-power-consuming, but expensive and insufficient. As the latest technique development trend, heterogeneous systems begin to emerge, providing high computing capacity, with much smaller cost. Field-programmable gate array (FPGA) and general purpose graphic processing units (GPGPU) (David et al. 2006) are used as synergistic processors, achieving high computing density and throughput.

Traditionally, GPU specifically works on graphic in computer systems. But modern GPGPU has been gradually moving towards general purpose computing. Other than CPU, GPGPU puts most of transistor budgets to processing unit. Only a few are used for controlling. So it is especially suitable for concurrency processing tasks with single instruction multiple data (SIMD) model. Old-fashion DSP, such as ADI's TS201, only has 3.6 GFLOPS (Qiang et al. 2009); mainstream quad-core CPU has about 40~50 GFLOPS (Victor et al. 2010); NVIDIA Tesla C2050 has 515 GFLOPS double-precision and 1.03 TFLOPS single-precision, with 448 processing cores and memory bandwidth is up to 144 GB/s; NVIDIA Tesla C2090 has 665 GFLOPS, 1.33 TFLOPS and 177 GB/s, respectively (NVIDIA Tesla C2050 Specification 2010). Those comparisons show that GPGPU system has high computing power. With efficient parallel programming model, it can handle heavy parallel tasks and show great potentials in large-scale signal processing field (Jinghong et al. 2009). Mccool, M.D. analyzed many application matters of using GPGPU for signal processing in paper (Mccool 2007); Blom and Follo (2005) used GPGPU to achieve high speed SAR imaging; by using GPGPU, Karasev et al. (2007) obtained 35× speedup on 2-dimensional phase unwrapping.

In this paper, based on GPGPU, high-speed multi-velocity-channel processing of Costas signal pulse compression is achieved, which provides a verification case and good reference for heterogeneous stream processing design for high speed radar signal processing.

10.2   Costas Signal Pulse Compression

10.2.1   Costas Echo Signal

Costas signal is a time-frequency hopping signal. Its carrier frequency varies with Costas code. It can be expressed as

$$g(t) = \frac{1}{\sqrt{NT_c}} \sum_{n=1}^{N} u_n \left(t - (n-1)T_c\right) \tag{10.1}$$

Among which,

$$u_n(t) = \text{rect}(t/T_c)\exp(j2\pi f_n t) \quad f_n = (c_n - 1)\Delta f \tag{10.2}$$

Therein, rect( ) is unit gating signal; T_c is sub-pulse width; N is sub-codes number, f_n is the frequency of the n-th pulse, c_n is Costas coding sequence; f is frequency hopping step. Generally, $\Delta f = 1/T_c$. The spectrum of above Costas signal is

$$G(f) = \sqrt{\frac{T_c}{N}} \sum_{n=1}^{N} \left\{ \sin c\left(T_c(f - f_n)\right) \times \exp\left(-j2\pi f(n-1)T_c\right) \right\} \tag{10.3}$$

For a target with a distance of r_0 and a constant radial velocity, the echo signal after coherent demodulation is

$$s(t) = g\left(\delta_{cm}(t - \tau_o)\right) \times \exp\left(-j2\pi f_c \tau_o\right) \times \exp\left(-j2\pi f_d(t - \tau_o)\right) \tag{10.4}$$

Therein, δ_{rm} and δ_{cm} are relative velocity factors and echo scale factors of the target respectively; f_d and $\tau_o = 2r_o/c$ are Doppler frequency and echo delay; c is light speed, f_c is carrier frequency; $\lambda = c/f_c$ is subcarrier wavelength. They have relations:

$$\delta_{rm} = 2v_r / c, \quad \delta_{cm} = 1 - \delta_{rm} \quad f_d = f_c \delta_{rm} \tag{10.5}$$

Generally, if bandwidth-time product of radial velocity and transmit signal satisfy $2v_r/c \ll 1/(BT)$, the baseband signal affected by scaling effect can be ignored.

Narrow band analysis model can be used for signal pulse compression. The echo signal spectrum approximates to

$$S(f) = G(f + f_d) \exp(-j2\pi(f_c + f)\tau_o)$$ (10.6)

where $G(f)$ is baseband Costas signal spectrum.

10.2.2 Computation Analysis

For narrow band analysis model, narrow ambiguity function can be used to analyze waveform characteristics. The one of Costas signal can be expressed as

$$\chi(\tau, f_d) = \int g(t) g^*(t+\tau) \exp(j2\pi f_d t) dt$$

$$= \frac{1}{N} \sum_{n=1}^{N} \exp(j2\pi(n-1)f_d T_c) \times \left\{ \Phi_{nn}(\tau, f_d) + \sum_{m=1, m \neq n}^{N} \Phi_{mn}(\tau - (m-n)T_c, f_d) \right\}$$ (10.7)

Among which,

$$\Phi_{mn}(\tau, f_d) = \operatorname{rect}\left(\frac{\tau}{2T_c}\right)\left(1 - \frac{|\tau|}{T_c}\right) \times \operatorname{sinc}(\alpha) \exp(-j\beta - j2\pi f_n \tau)$$ (10.8)

and,

$$\alpha = (f_m - f_n - f_d)(T_c - |\tau|), \quad \beta = \pi(f_m - f_n - f_d)(T_c + |\tau|)$$ (10.9)

The ambiguity function approximately shows "thumbtack" shape. It could provide better distance and velocity resolution, while maintaining lower side-lobe characteristics. It is an ideal waveform.

For Costas frequency modulation signal, multi-channel processing is needed to get target information about distance and radial velocity. If transmitted signal duration is T_p, as for the maximum radial velocity of target is V_{rmax}, the Doppler channel number is (V_{rmax} integer of upward taken)

$$N_d = 2\left\lceil \frac{2v_{rmax}/\lambda}{1/T_p} \right\rceil = 2\left\lceil \frac{2T_p v_{rmax}}{\lambda} \right\rceil$$ (10.10)

Figure 10.1 shows the multi-channel processing. The main operation in pulse compression process of Costas signal is the fast Fourier transform (FFT). Considering the reference spectrum can be pre-calculated and stored, its computing time can be negligible. Hence the most computing time is consumed on FFTs. For FFT

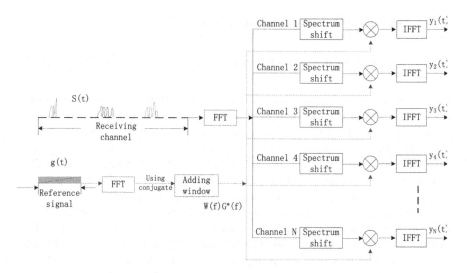

Fig. 10.1 Costas signal multi-channel pulse compression diagram

operation with length of P (P is of 2's integer power), the butterfly computation number is $(P/2)\log_2 P$.

Assuming the transmitting signal pulse width is T_p, bandwidth is B, sampling frequency is f_s, maximum range is R_M, the duration of receiving channel is $T_P + 2R_M/c$. Accordingly, receiving channel sampling points P is 2^M, among which, $M = \lceil \log_2((T_p + 2R_M/c)f_s) \rceil$. In conclusion, the calculation time of multi-channel pulse compression T_{pc} can be expressed as

$$T_{pc} = N_d \cdot \left(0.5 T_B P \log_2 P + P T_M\right) \approx N_d \cdot \left(0.5M + 1\right) \cdot P \cdot T_M \qquad (10.11)$$

in which, T_B and T_M are the computing times of butterfly computation and complex multiplication computation, respectively. It is commonly assumed that $T_B \approx T_M$.

Take America "pave paws" AN/FPS-115 radar as an example, it is of maximum pulse width 16 ms; maximum signal width 1 MHz, max effective distance 6,000 km. If two-channel I/O sampling frequency is 4 MHz, according to (10.11), the range gate number is 262,144. Combined with (10.10), N_d=5,418 Doppler processing channels are needed. $M = \lceil \log_2((T_p + 2R_M/c)f_s) \rceil = 18$. If the time interval of the pulse compression operation is 10 ms, the total complex multiplication number is:

$$N_d \cdot \left(0.5M + 1\right) \cdot P \cdot 5 = 5418 \cdot \left(0.5 \cdot 18 + 1\right) \cdot 2^{18} \cdot 5 \approx 7.1 \times 10^9 \qquad (10.12)$$

This is divided by 10 ms to get the computing requirement 7.1 TFLOPS. If fully using DSP to build the processing system, the cost and difficulty will be huge. In fact, putting the commercial products in use has become priority consideration in long range warning radar and other high performance radar processing systems (Derham et al. 2003).

10.3   GPGPU Processing System Design

10.3.1   CPU/GPGPU Heterogeneous Supercomputing Platform

Heterogeneous computing platform is a processing system built by different types of processors. With the development of integrated circuit technique, the traditional model of main frequency and single processing core improvement has been constrained by the physical limit. Multi-core systems become the new trend. Commonly, heterogeneous system is made of general processing cores and massive parallel computing cores: the former complete the task of logic control, data management, and user interaction and so on; the latter is responsible for processing high density computing tasks. In signal processing field, FPGA and DSP mixed structure build the typical heterogeneous systems. However, CPU-GPU heterogeneous system has superior processing capacity, while keeping scalability and flexibility, which leads to a very nice signal processing platform. Figure 10.2 shows the latest hardware architecture of CPU/GPGPU heterogeneous system. CPU and GPGPU are connected through high-speed inter-node bus within the same node and then interconnected by network between nodes. CPU is responsible for task flow control, data distribution, preprocessing and part of computing, while GPGPU handles the heavy computing tasks, such as FFT or matrix operations.

10.3.2   Software Task Flow Design

Through the analysis of Chap. 2, we can map the Fig. 10.1 into hardware as shown in Fig. 10.3.

Firstly, CPU main controller starts overall processing. It transfers data to GPGPU. GPGPU then launches many hardware threads, executes the parallel algorithms, and

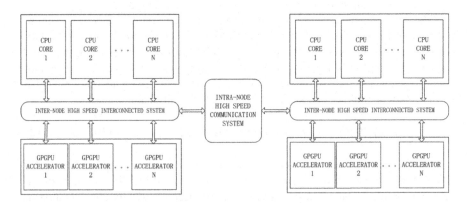

Fig. 10.2 Hardware architecture of heterogeneous CPU/GPGPU processing system

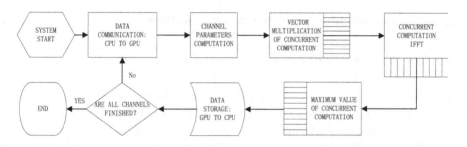

Fig. 10.3 Flow graph of task

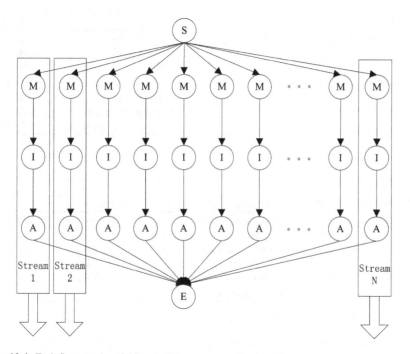

Fig. 10.4 Task flow graph unfolding and stream processing model

then transfers the final results back. Different channels processing are independent, so effective loop unrolling could be obtained. The data flow diagram is presented in Fig. 10.4. The "S" state means startup, which prepares all the needed resources and data. "M" state means point-to-point multiplication and "I" state performs IFFT operation. The "A" state computes the maximum. Every branch could be organized into a GPU stream, which could be scheduled and optimized with overlapping data transfer and execution. Finally, all the results are paralleled reduced to "E" end state, where only one result is produced.

10.4 Experimental Verification and Analysis

10.4.1 Experimental Environment and Parameters

Experiments imitate "pave paws" radar parameters. The tests are performed on an 8-GPU accelerated computing server. Table 10.1 lists the test parameters; Table 10.2 lists software configuration; Table 10.3 contains hardware specifications. Through GPGPU, the hardware provides a peak computing power of more than 8TFlops.

Table 10.1 Test parameter

Parameter	Value
Carrier frequency	4 GHz
Band width	1 MHz
Sampling frequency	4 MHz
Maximum distance	6,000 km
Pulse width	16 ms
Doppler channel	5,418
Range gate unit number	262,144

Table 10.2 Test software environment

Software	Version
Operating system	Redhat Linux AS 5.5 64bit
NVIDIA CUDA library version	3.2
NVCC compiler	3.2 64bit
cuFFT library	3.2 64bit
cuBlas library	3.2 64bit
gcc compiler	4.3

Table 10.3 Hardware environment

Hardware	Type	Value
CPU	Intel Nehalem 552 2.27 GHz	2
Memory	DDR3 1,033 MHz	36GB
GPU accelerator	Tesla C2050 448-core	1
GPU accelerator	Tesla C1060 240-core	8
GPU video memory	GDDR3	32GB
GPU video memory	GDDR5	3GB

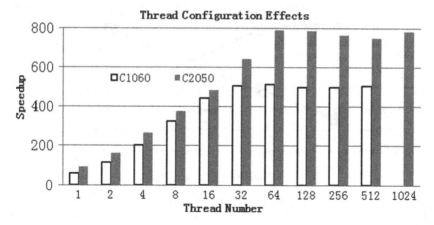

Fig. 10.5 Block internal threads number effect on performance

10.4.2   Experimental Results

1. Threads configuration
   Firstly, the effects on different concurrent threads within single device are tested.
   For comparison, relative throughput is used and expressed as the processing time
   divided by 105. The speedup against the Intel Nehalem E5520 CPU is shown in
   Fig. 10.5. For the Tesla C1060 series, because each internal execution unit can
   simultaneously launch 32 hardware threads, maximum throughput is achieved
   after every block has 32 threads. For C2050, this number is 64. The tests show
   that when fully occupancy is achieved, it is not always true to get better performance with more hardware threads.
2. Speedup and Efficiency
   Figure 10.6 shows the speedup and efficiency obtained from using different
   number of accelerators. More accelerators lead to larger speedup. But the growth
   rate of speedup slows down. The efficiency reduces with the number of accelerator increases. It means that other overheads (such as scheduling, queuing and
   communication) increase with the system scale, which requires more optimization. But the final efficiency reaches more than 50 %.
3. Final Results
   A high-end dual-slot server is taken as test bed. It has two Intel Nehalem CPU @
   2.27 GHz with totally eight cores. One Tesla C1060 can get over 150 times
   speedup. When using 8 C1060, 574× speedup is achieved. The processing time
   is reduced to nearly 0.5 s. But if real-time processing is required, over ten times
   computing power of this platform is required. The real-time processing makes a
   great challenge (Table 10.4).

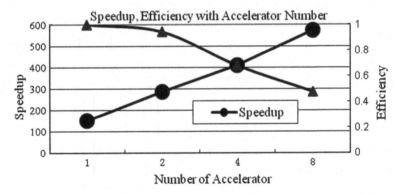

Fig. 10.6 Speedup and efficiency varies with accelerators number

Table 10.4 Final optimization results

Platform	Execution time (s)	Speedup
CPU Intel Nehalem	295.24	1
GPU C1060	1.956	151
GPU C2050	1.265	233
GPU C1060*2	1.036	285
GPU C1060*4	0.719	411
GPU C1060*8	0.514	574

10.5 Summary and Future Work

In this paper, through the analysis of large-scale pulse compression for Costas signal, a parallel processing scheme is proposed. The same parameters with America "pave paws" radar are used and tested on heterogeneous CPU/GPGPU computing system. The experimental results show that, the GPGPU accelerated system has supercomputing capacity, and can get higher performance speedup than general CPU system. The test system with 8 GPGPU accelerators can obtain 574 times speedup and processing time of 514.3 ms. It is proved that GPGPU accelerated system can effectively deal with large-scale radar signal processing tasks. The future work includes fully real-time processing system design, scheduling optimization and the distributed system design.

Acknowledgements Dr. Bin ZHOU works as the chief scientist of marine remote sensing lab at Institute of Oceanographic Instrumentation, Shandong Academy of Sciences. This work is supported by the institute funding.

References

Blom, M., Follo, P., 2005, VHF SAR image formation implemented on a GPU. Geoscience and Remote Sensing Symposium. IGARSS '05. Proceedings. 2005 IEEE International, pp: 3352 – 3356.

David L, Mark H, Naga G, et al., 2006, GPGPU: general-purpose computation on graphics hardware. Proceedings of the 2006 ACM/IEEE conference on Supercomputing, 208.

Derham, T., Woodbridge, K., Griffiths, H., et al., 2003, The design and development of an experimental netted radar system. Proceedings of the International Radar Conference, 293–298.

Hongxun, H., Xiuchun, W., 2004, Application of Parallel Computer to Modern Radar Signal Processing. Morden Radar, 26(3):25–32.

Jiang Jinghong, J., Li, Y., Huizhi, C., Weiming, H., 2009, Feasibility research of real-time signal processing system using GPU. Technical Acoustics, 28(2):129–131.

Karasev P.A., Campbell D.P., Richards M.A., 2007, Obtaining a 35x Speedup in 2D Phase Unwrapping Using Commodity Graphics Processors. IEEE Radar Conference, Boston, MA, 574–578.

Mccool, M.D., 2007, Signal Processing and General-Purpose Computing and GPUs [Exploratory DSP]. IEEE Signal Processing Magazine, 24(3):109–114.

NVIDIA Tesla C2050 Specification, 2010, http://www.nvidia.com/object/product_tesla_C2050_C2070_us.html.

Victor W.L., Changkyu, Chhugani, J., et al., 2010, Debunking the 100X GPU vs. CPU myth: an evaluation of throughput computing on CPU and GPU. ISCA '10 Proceedings of the 37th annual international symposium on Computer architecture, 38(3):451–460.

Qiang, W., Qing, G., Xuwen, L., 2009, Hardware design of image information processor based on ADSP-TS201 DSPs. 2009 IEEE International Workshop on Imaging Systems and Techniques, 155–158.

Wenhua, M., Yixian, Y., 1996, Frequency hopping encoding technique in morden radar. Morden Radar, 18(5):82–88.

Zhiguang, X., Yingmin, W., Zhiqiang, L., 2010, Analysis of detective performance of sonar target based on Gostas signal. Audio Engineering, 34(3):49–53.

Chapter 11
Parallelizing ISODATA Algorithm for Unsupervised Image Classification on GPU

Fei Ye and Xuan Shi

Abstract Iterative Self-Organizing Data Analysis Technique Algorithm (ISODATA) is commonly used for unsupervised image classification in remote sensing applications. Although parallelized approaches were explored, previous works mostly utilized the power of CPU clusters. We deploy the many-cores in the Graphics Processing Unit (GPU) to accelerate the unsupervised image classification over GPU. The proposed solution is scalable and satisfactory to speed up the computational time, while the quality of classification is almost the same as that from ERDAS, a well known remote sensing software.

Keywords ISODATA • Unsupervised image classification • Graphics processing unit (GPU)

11.1 Introduction

Graphics Processing Unit (GPU) was traditionally built for the purpose of efficiently manipulating computer graphics. For this reason, image processing has been a primer application that benefit from the marvelous parallelism inherent in graphic processing. Today's GPU has been rapidly evolved to support General-Purpose computing [thus called GPGPU] though many-core processors capable of intensive computation and data throughput thus a modern GPU enables massively parallel computing and is not limited only to process computer graphics. By simultaneously

F. Ye
School of Computational Science and Engineering, Georgia Institute of Technology, 266 Ferst Dr. NW, Atlanta, GA 30332, USA
e-mail: yefei@gatech.edu

X. Shi (✉)
Department of Geosciences, University of Arkansas, Fayetteville, AR 72701, USA
e-mail: xuanshi@uark.edu

X. Shi et al. (eds.), *Modern Accelerator Technologies for Geographic Information Science*,   145
DOI 10.1007/978-1-4614-8745-6_11, © Springer Science+Business Media New York 2013

executing tens of thousands of threads from hundreds of specialized GPU cores, GPGPU can even achieve high performance computing over the desktop and laptop computers (Pankratius et al. 2011).

While satellite imagery and aerial photo are increasingly available, processing quantum size of data needs efficient solutions to complete the tasks of image analysis. Accelerating the computation to improve the performance of image processing and analytics will directly benefit the societal needs. Among a variety of image processing modules, unsupervised image classification has been fundamental but computation intensive. This paper introduces the implementation of a classic algorithm for unsupervised image classification over the GPU. Iterative Self-Organizing Data Analysis Technique Algorithm (ISODATA) is a heuristic procedure that automatically classifies the image through multiple iterations for optimization. In comparison to the same function implemented over a commercial software product, the solution on the GPU can achieve a speedup up to 40+ depending on the number of classes. In face of the big data challenge in image processing, deploying multiple GPUs over hybrid computer architectures and systems will be the future direction to achieve both scalability and high performance.

11.2 Parallelizing ISODADA for Unsupervised Image Classification

ISODATA is one of the most frequently used algorithms for unsupervised image classification algorithms in remote sensing applications. It is also the default algorithm for unsupervised image classification in ERDAS, a well-known remote sensing software. This section will first summarize the ISODATA algorithm and its implementation in sequential procedures. Implementing the parallelized solution on CUDA/GPU is followed by a brief review of previous works on parallelized unsupervised image classification.

11.2.1 ISODATA in Sequential Procedure

In general, ISODATA can be implemented in three steps: (1) calculate the initial mean value of each class; (2) classify each pixel to the nearest class; and (3) calculate the new class means based on all pixels in one class. The second and third steps are repeated until the change between two iterations is small enough.

To perform ISODATA classification, several parameters need to be specified in Table 11.1.

In ERDAS and this pilot study, the initial class means is derived by the statistics of original data sets, although the initial means can be assigned arbitrarily. Accordingly, the initial class means are evenly distributed in a multi-dimensional

Table 11.1 Prerequisites for ISODATA implementation

Symbol	Definition
C	The number of classes to be created
T	Convergence threshold which is the maximum percentage of pixels whose class values are allowed to be unchanged between iterations
M	The maximum number of iterations to be performed

feature space along a vector between two points $(\mu_1 - \sigma_1, \mu_2 - \sigma_2, \ldots, \mu_k - \sigma_k)$ and $(\mu_1 + \sigma_1, \mu_2 + \sigma_2, \ldots, \mu_k + \sigma_k)$ in the first iteration, where μ_i denotes the mean of the ith band ($i = 1, 2, \ldots, k$) and k is total number of bands in the dataset, and σ_i is the standard deviation of band i. In the following iterations, when the maximum number of iterations (M) and the convergence threshold (T) are not reached, the means of all classes are recalculated, causing the class means to shift in the feature space. During the iterative calculations, each pixel is compared to the new class means and assigned to the nearest class mean.

During the process of classification, each class is labeled as a certain type of object. The change between two consecutive iterations can be either the percentage of pixels whose class labels have been changed between two iterations, or the accumulated distances of the class means that have been changed in the feature space between two iterations. The iterative process will be terminated until the convergence threshold (T) is reached or exceeded, which means the change between two iterations is *small* enough, that is, the *maximum* percentage of pixels whose class values that are *unchanged* between iterations. Classic ISODATA algorithm also includes merging and splitting of classes between iterations for refinements. Since ERDAS is used to compare the quality and performance of the proposed solution, this paper does not consider the refinement issues since ERDAS does not include these features.

The sequential ISODATA algorithm can be described in the following pseudo-code:

```
procedure label := ISODATA (data, C, M, N, T)
/* data is the input image dataset */
/* C is the number of classes to be created */
/* M is the maximum number of iterations */
/* N is the number of pixels */
/* T is the convergence threshold*/
/* label is the returned array containing the label of
class to which each pixel is assigned */
begin
   Calculate mean of the value of the pixels in the data-
   set (mean);
   Calculate standard deviation of the value of the pixels
   in the dataset (dev);
   Calculate initial means of C classes (class_mean);
```

```
m := 0; /* m is the counter of the iteration */
con := 0;   /* con is convergence*/
while m < M AND con < T
  Update class means based on the pixels in one class
  (class_mean);
  Calculate the labels (label);
  Count the labels of each class (count);
  Calculate the convergence (con);
  Update the previous labels (pre_label);
  m := m + 1;
end while
end ISODATA
```

In summary, the procedures will be iterated until the two preconditions ($m < M$ and $con < T$) are satisfied.

11.2.2 Previous Works on Parallelized Unsupervised Image Classification

Obviously when more iterations are executed, it takes longer time to complete the classification tasks. In order to gain speedup, several approaches were reported by deploying MapReduce and clusters of computers or workstations via Message Passing Interface (MPI).

MapReduce (Dean et al. 2004) is a programming model introduced by Google in 2004 to support distributed computing on large data sets on clusters of computers. Zhao et al. (2009) proposed a parallel k-means clustering algorithm based on MapReduce. It experimented on 1GB, 2GB, 4GB and 8GB datasets on cluster of 1, 2, 3 and 4 nodes. The limitation of this research is the resulted speedup is around 3 which might not be very satisfactory. Inspired by the this work, Lv et al. (2010) implemented a parallel K-means clustering of remote sensing images based on MapReduce which has a speedup of around 10.

Li et al. developed a parallel ISODATA algorithm in MapReduce (Li et al. 2010). Its environment setting was 1 NameNode and 8 DataNodes. The NameNode used 8 2.0GHz CPUs and 4GB RAM. Each DataNode used 8 2.0GHz CPUs and 8GB RAM. All nodes shared 20TB drive space and were connected by a gigabit switch. Its experiment was performed on images of data size of 1GB, 2GB, 4GB and 8GB respectively. It did show some execution time speedup. But the limitation of this research is that it didn't take into account the condition of only 1 node. The speedup of 8 nodes versus 4 nodes is roughly 5. In general, converting binary image data into ASCII text files to use MapReduce seems not an optimized and practical solution in remote sensing applications.

Message Passing Interface (MPI) (Snir et al. 1995) is a standard for parallel programs running on computer clusters or supercomputers by sending and receiving messages to communicate between computing processes. Riccardi and Schow implemented ISODATA on a vector computer and it ran about ten times faster than its corresponding scalar version (Riccardi 1988). However, this old fashioned supercomputer is no longer competitive. In another research, a distributed ISODATA (D-ISODATA) algorithm was designed and developed (Dhodhi et al. 1999). It proposed a supervisor-worker mode to use a supervisor node to coordinate other workers nodes over a network of workstations. The basic idea of this approach is to distribute time-consuming part of the algorithm among workers. It took calculation of the Euclidean distance as the time-consuming part. By dividing image pixels into smaller and equal sized blocks, each worker as well as the supervisor is responsible for computation of pixels in its own assigned block. When running on a network of eight workstations, it resulted in an execution time speedup of less than 8.

11.2.3   ISODATA by CUDA/GPU

Compute Unified Device Architecture (CUDA) is NVIDIA's general-purpose parallel computing architecture. Here, the Central Processing Unit (CPU) is referred to as a host, while an individual GPU is referred to as a device. The kernel is the function that runs on the device and is executed by an array of threads, while all threads can run the same code concurrently. Each thread has a unique thread identifier and can be accessed via the threadIdx variable. Thread identifiers (threadIDs) can be defined in one, two or three dimensions. Furthermore, threads can be grouped into thread blocks and grids. Threads in same thread block can cooperate with each other via shared memory, atomic operations or barrier synchronization. Threads in different blocks cannot cooperate. A user-defined number of threads can be organized in a block with a maximum number of 512 threads. Similarly a group of thread blocks can be organized into a grid in which each thread may be executed independently and thus may execute in parallel.

In this pilot study, the GPU is a NVIDIA GeForce GTS 450, which has 192 cores. According to the technical specification, this GPU has 24 streaming multiprocessors (SM). Each SM has eight CUDA cores called as streaming processor (SP). In this GTS 450 with a compute capability of 2.1, up to 1,024 threads can be assigned to each SM. Thus a maximum of $1,024 \times 24 = 24,576$ threads can run concurrently in parallel on the physical GPU, although the maximum sizes of each dimension of a block is $512 \times 512 \times 64$ and the maximum sizes of each dimension of a grid is $65,535 \times 65,535 \times 1$. If the number of threads is more than the maximum number [24,576], the remaining threads have to wait.

In the CUDA programming model, ISODATA algorithm can be parallelized in many procedures documented in the above sequential pseudo-code: (1) The computational procedures of the mean value and standard deviation can be parallelized since they are basically the summation over all pixels in the whole dataset; (2) After the first iteration, the computation of class means can also be parallelized since it consists of summations over pixels in each class; (3) The labeling process, which is assigning class labels to all pixels, can be carried out in parallel; (4) Operations for counting the labels in each class and the calculation of convergence can be parallelized; and (5) the procedures for updating class labels of the previous iteration can be easily parallelized.

These five processes can be parallelized and are *highlighted* in the following pseudo-code that can be conveniently implemented in CUDA programming in which the computing kernels have ample amount of parallelism:

```
procedure label := ISODATA_CUDA (data, C, M, N, T)
/* data is the input image dataset */
/* C is the number of classes to be created */
/* M is the maximum number of iterations */
/* N is the number of pixels */
/* T is the convergence threshold*/
/* label is the returned array containing the label of
class to which each pixel is assigned */

begin
  /* calculate mean of the dataset - parallelized pro-
  cess (1)*/
  mean := PARALLEL_REDUCTION_MEAN (data);
  mean := mean / N;
  /* calculate standard deviation of the dataset - par-
  allelized process (1)*/
  dev := PARALLEL_REDUCTION_DEV (data);
  dev := dev / N;
  /* calculate initial means of C classes */
  step := dev * 2 / (C - 1);
  for i := 0 to C - 1
    class_mean[i] := mean - dev + i * step;
  end for
  m := 0; /* m is the counter of the iteration */
  con := 0;   /* con is convergence*/
  while m < M AND con < T
    /* update class means based on pixels in one class -
    parallelized process (2)*/
    if  m != 0 then
        for i := 0 to C - 1
          class_mean[i] := PARALLEL_REDUCTION_CLASSMEAN
        (data, label, i);
          class_mean[i] := class_mean / count[i];
        end for
    end if
  /* calculate labels - parallelized process (3)*/
  label := PARALLEL_CAL_LABEL(data, class_mean);
  /* count labels of each class - parallelized process
  (4) */
```

```
for i := 0 to C - 1
  count[i] := PARALLEL_REDUCTION_COUNT (label, i);
end for
/* calculate convergence - parallelized process (4)*/
if m != 0 then
  con := PARALLEL_REDUCTION_SAME (pre_label, label);
  con := con / N;
end if
/* update previous labels - parallelized process (5)*/
pre_label := PARALLEL_PRELABEL (label);
m := m + 1;
end while
end ISODATA_CUDA
```

All CUDA functions in the above pseudo code are marked with *PARALLEL* at the beginning, such as *PARALLEL_REDUCTION_MEAN*. Below is a CUDA code segment that implements this function to calculate the mean value of the dataset.

```
// PARALLEL_REDUCTION_MEAN is a reduction (summation)
function on the GPU for calculating the mean
// value of the input image dataset. Below is a description
about the input and output variables used in this
function.
// Input: fDataset_d - data stored in device memory
// sdata_d - memory used to store intermediate results
// nNumData - number of data (data size)
// nNumBands - number of bands (layers)
// k - current band number for which we are calculating
the sum
// Output: odata_d - the output of each block in device
memory
__global__ void reduce_mean(float *fDataset_d, float
*odata_d, float *sdata_d, int nNumData, int k)
{
  unsigned int tid = threadIdx.x; // get thread indux
  unsigned int i = blockIdx.x*blockDim.x + threadIdx.x;
// get global index
  sdata_d[i] = (i < nNumData) ? (fDataset_d[k*nNumData +
i]) : 0.0; // load data from device memory
  __syncthreads();
  // do reduction in global mem
  for(unsigned int s = blockDim.x / 2; s > 0; s >>= 1) {
    if (tid < s) {
      sdata_d[i] += sdata_d[i + s];
    }
    __syncthreads();
  }
```

```
// write result for this block to global mem
if (tid == 0) odata_d[blockIdx.x] = sdata_d[blockIdx.x*
blockDim.x + 0];
}
```

In CUDA, a function using the __global__ declaration specifier is called a *kernel* or *device* function and runs on the device. A function using __host__ declaration specifier is called the host function and runs on the host. Parallel portions of a program are executed on the device as kernels. The keywords "threadIdx.x", "blockIdx.x", and "blockDim.x" in the above CUDA code refer to the index of a thread in a certain thread block that has a maximum of three dimensions. Since all threads execute the same kernel function, CUDA API offers such a mechanism to distinguish each thread by a unique identifier through a combination of the thread identifier, the block identifier, and the dimension of the block. Thus a thread can be referred by blockIdx.x*blockDim.x + threadIdx.x.

Since the host (CPU) and devices (GPU) have separate memory spaces, in order to execute a kernel function on a device, the CUDA program needs to first allocate the memory for both input and output data on the device and transfer the data from the host to devices. Similarly, after the execution is completed on the device, the output data needs to be transferred from the device back to the host and free up the device memory that is no longer used.

Given the example in the above mean calculation, before this function is implemented, the data [named as *fDataset*] on the CPU memory has to be transferred to the GPU memory [named as *fDataset_d*] through the following CUDA API *cudaMemcpy*. The enumeration variable *cudaMemcpyHostToDevice* denotes that this operation transfer the data from the host memory to the device memory by assigning the size of the memory as *sizeof(float)*nXSize*nYSize*nNumBands*. Here on the device memory, the data type is float. The size is the number of pixel rows times the number of pixel columns times the number of the bands, while such information is derived from GDAL API when reading the input image.

```
cudaMemcpy ( fDataset_d, fDataset, sizeof(float)*nXSize*
nYSize*nNumBands, cudaMemcpyHostToDevice);
```

Given the other example, after the classification is completed, the output data [named as *nLabel_d*] on the device memory has to be transferred back to the host memory [named as *nLabel*] through the CUDA API *cudaMemcpy*. The enumeration variable cudaMemcpyDeviceToHost denotes that this operation transfer the data from the device memory to the host memory by assigning the data type as *unsigned char* and the data size as *sizeof(unsigned char)*nXSize*nYSize*.

```
cudaMemcpy(nLabel,      nLabel_d,      sizeof(unsigned
char)*nXSize*nYSize, cudaMemcpyDeviceToHost);
```

To invoke the above kernel function from the CPU to calculate the mean value of the input image dataset, CUDA API has a special syntax to specifies the dimensions of the block per grid and the dimensions of each thread block between <<< and >>>, such as

```
reduce_mean    <<<    blocksPerGrid,    threadsPerBlock    >>>
(fDataset_d, odata_d, sdata_d, nNumData, k);
```

This kernel function is called from a host function to do the summation of pixel values in each band of the image data by utilizing such a series of threads for parallel computing.

11.3 Comparison of the Quality and Performance of CUDA Program

The proposed algorithm is implemented in CUDA C program and compiled in Visual Studio 2010 with CUDA Toolkit 3.2. The program runs on a single desktop computer, in which the CPU is Intel Pentium 4 CPU with 3.00 GHz main frequency, while the RAM size is 4 GB. The GPU on this machine is GeForce GTS 450 which has 192 cores with the Fermi architecture and has 1 GB global memory. The image data used in this pilot study, dc_ikonos_subset.img, can be found from ERDAS data DVD. The file size is 89.6 MB and has three bands. The image has 15.6 million pixels which are 4,293 pixels in width and 3,263 pixels in height. Data I/O streaming is implemented via GDAL APIs. Figure 11.1 displays the source image, while Fig. 11.2 displays the result of classification.

The proposed solution of parallelized ISODATA on GPU is evaluated by comparing both the quality and the performance to the same unsupervised classification function in ERDAS 2010. Table 11.2 describes the convergence values derived from CUDA and ERDAS in the case when the image is classified into five classes ($C = 5$). Table 11.3 displays the number of pixels in each class in the case when the image is classified into five classes ($C = 5$).

The performance is evaluated by the execution time of ISODATA by CUDA and by ERDAS when the image is classified into different classes. The speedup can be defined as $Speedup = Te/Tc$, where T_c is the execution time by CUDA and T_e is the execution time by ERDAS. In this experiment, T_c includes the time of reading image file, classification, and writing output result. T_e also includes file reading, classification and writing output result but does not include the time of signature accumulation. Table 11.4 is a comparison of the execution time and number of iterations by CUDA and ERDAS when the image is classified into different classes.

11.4 Conclusion

This paper proposed a method to parallelize ISODATA algorithm for unsupervised image classification using CUDA on the GPU. The parallel execution of ISODATA on GPU achieves much better performance while the classification results have desirable quality. Thus the high performance computing power of GPUs is verified in processing remote sensing image data. How to deal with the memory limit on the GPU to process large scale image data remains a significant research challenge in

Fig. 11.1 The source image

Fig. 11.2 The result of classification

Table 11.2 Comparison of the convergence value of each iteration by CUDA and ERDAS ($C=5$)

Iteration	Convergence value by CUDA	Convergence value by ERDAS
1	0.000	0.000
2	0.910	0.910
3	0.943	0.943
4	0.948	0.948
5	0.950	0.950
6	0.951	>=0.95

Table 11.3 Comparison of number of pixels in each class by CUDA and ERDAS ($C=5$)

Class	# of pixels in each class by CUDA	# of pixels in each class by ERDAS
1	3,808,329	3,808,329
2	4,038,247	4,038,270
3	4,303,206	4,303,184
4	2,907,823	2,907,825
5	495,934	495,931

Table 11.4 Performance comparison

Classes	CUDA		ERDAS		
	Time (Tc)	# of iteration	Time (Te)	# of iteration	Speedup times (Te/Tc)
3	3.602	4	109.2	4	30
4	4.513	6	188.6	6	42
5	5.430	6	227.4	6	42
6	4.274	4	149.6	4	35
7	4.692	4	111.7	4	24
8	5.316	4	82.5	4	16
9	4.966	4	60.6	4	12
10	5.260	4	54.7	4	10
11	6.062	4	52.1	4	9
12	5.577	4	45.8	4	8
13	7.868	4	48.6	4	6
14	6.005	4	50.5	4	8
15	6.306	4	49.2	4	8
20	15.079	9	120.8	9	8

the future. Currently, GPU for supercomputing use may have up to 6 GB memory which might not be sufficient enough to process large image data that has high resolution and large file size that is over several gigabytes, considering the memory needs to handle intermediate outcomes from parallelized computation. Therefore, a memory swapping strategy needs to be explored when a single GPU is used.

Alternatively large image data can be processed in parallel on CUDA by distributing the data and computation onto multiple GPUs. This solution can be implemented on supercomputers which have multiple GPUs and CPUs using MPI to schedule tasks between CPUs and GPUs.

Acknowledgements This research was supported partially by the National Science Foundation through the award OCI-1047916.

References

Bo Li; Hui Zhao; ZhenHua Lv. 2010. Parallel ISODATA Clustering of Remote Sensing Images Based on MapReduce. In *Cyber-Enabled Distributed Computing and Knowledge Discovery (CyberC), 2010 International Conference on*, pp. 380-383, 10–12 Oct. 2010. doi: 10.1109/CyberC.2010.75.

Dean, Jeffrey and Ghemawat, Sanjay. 2004. MapReduce: Simplified Data Processing on Large Clusters. OSDI'04: Sixth Symposium on Operating System Design and Implementation, 2004. http://labs.google.com/papers/mapreduce-osdi04.pdf

Dhodhi M.K., Saghri J.A., Ahmad I., Ul-Mustafa R. 1999. D-ISODATA: A Distributed Algorithm for Unsupervised Classification of Remotely Sensed Data on Network of Workstations. Journal of Parallel and Distributed Computing, 59 (2), pp. 280–301.

Riccardi, Schow. 1988. Adaptation of the ISODATA clustering algorithm for vector supercomputer execution. Proceedings of the 1988 ACM/IEEE conference on Supercomputing vol. 2, 1988 pp.141–150

Snir, Marc; Otto, Steve; Huss-Lederman, Steven; Walker, David; Dongarra, Jack. 1995. MPI: The Complete Reference. MIT Press Cambridge.

Victor Pankratius, Wolfram Schulte, and Kurt Keutzer. 2011. Guest Editors' Introduction: Parallelism on the Desktop. IEEE Software, vol. 28, no. 1, pp. 14–16, Jan./Feb. 2011.

Weizhong Zhao, Huifang Ma, Qing He. 2009. Parallel K-Means Clustering Based on MapReduce. CloudCom '09 Proceedings of the 1st International Conference on Cloud Computing.

Zhenhua Lv, Yingjie Hu , Haidong Zhong, Jianping Wu, Bo Li, Hui Zhao. 2010. Parallel K-means clustering of remote sensing images based on mapreduce. Proceedings of the 2010 international conference on Web information systems and mining, October 23–24, 2010, Sanya, China

Chapter 12
Accelerating Mean Shift Segmentation Algorithm on Hybrid CPU/GPU Platforms

Miaoqing Huang, Liang Men, and Chenggang Lai

Abstract Image segmentation is a very important step in many GIS applications. Mean shift is an advanced and versatile technique for clustering-based segmentation, and is favored in many cases because it is non-parametric. However, mean shift is very computationally intensive compared with other simple methods such as k-means. In this work, we present a hybrid design of mean shift algorithm on a computer platform consisting of both CPUs and GPUs. By taking advantages of the massive parallelism and the advanced memory hierarchy on Nvidia's Fermi GPU and Kepler GPU, the hybrid design achieves a 30 ×speedup compared with the pure CPU implementation on the filtering step when dealing with images bigger than 4096 ×4096 pixels.

12.1 Introduction

Segmentation is the process of partitioning a digital image into multiple segments to simplify and/or change the representation of an image into a fashion that is more meaningful and easier to analyze. Image segmentation is typically used to locate objects and boundaries (lines, curves, etc.) in images (Shapiro and Stockman 2001). The result of image segmentation is a set of segments that collectively cover the entire image, or a set of contours extracted from the image. Many methods have been designed to achieve the image segmentation goal, such as thresholding, clustering, edge detection, region growing among others.

In this work, we focus on accelerating the mean shift image segmentation algorithm on hybrid CPU/GPU computer platforms. Mean shift was introduced by Fukunaga and Hostetler (1975) and has been extended to be applicable in other

M. Huang (✉) • L. Men • C. Lai
University of Arkansas, 1125 W Maple St, Fayetteville, AR 72701, USA
e-mail: mqhuang@uark.edu

X. Shi et al. (eds.), *Modern Accelerator Technologies for Geographic Information Science*, 157
DOI 10.1007/978-1-4614-8745-6_12, © Springer Science+Business Media New York 2013

fields like Computer Vision (Cheng 1995; Comaniciu and Meer 2002). Mean shift
is a versatile non-parametric iterative algorithm that has better robustness than other
clustering-based image segmentation algorithms such as k-means (Kanungo
et al. 2002). However, mean shift algorithm is computationally intensive. It's time
complexity is $O(T n^2)$, where T is the number of iterations for processing each data
point and n is the total number of data points in the data set, e.g., the number of
pixels in the image. As it is shown in Sect. 12.4, the processing time of a large image
becomes prohibitively long. This prompts the use of graphics processing units to
accelerate the most computationally intensive part of the mean shift algorithm.
Although the processing of an individual pixel is an iterative process, which is suitable to execute on CPU, the processing of individual pixels is independent. The
processing independence between individual pixels provides a good chance to
improve the performance by concurrently processing hundreds of pixels on massively parallel GPUs.

The remainder of the text is organized as follows. In Sect. 12.2 the mean shift
algorithm is briefly introduced. The implementation detail is described in Sect. 12.3.
Section 12.4 presents the results, and the conclusions are given in Sect. 12.5.

12.2 Mean Shift Algorithm

Mean shift considers feature space as an empirical probability density function.
If the input is a set of points then mean shift considers them as sampled from the
underlying probability density function. If denser regions are present in the feature
space, then they correspond to the local maxima of the probability density function.
For each data point, mean shift associates it with the nearby peak of the data set's
probability density function. For each data point, mean shift defines a window
around it and computes the mean of the data point. Then it shifts the center of the
window to the mean and repeats the algorithm till it converges. After each iteration,
we can consider that the window shifts to a more denser region of the data set. At
the high level, we can specify mean shift as follows:

1. Define a window around each data point;
2. Compute the mean of data within the window;
3. Shift the center of the window to the mean and repeat till convergence, i.e., the
   center of the window no longer shifts.

This process is illustrated in Fig. 12.1 in which particles are used as an example.
The original position of the window is shown in Fig. 12.1a. The geometric center of
the window (i.e., $G\ C_1$) does not overlap with the center of the mass (i.e., $M\ C_1$).
Therefore, the geometric center of the window is shifted to the center of the mass
(i.e., $G\ C_2 \leftarrow M\ C_1$) in Fig. 12.1b. Then a new center of the mass is calculated (i.e.,
$M\ C_2$). If the geometric center and the mass center do not overlap, the geometric
center will keep shifting until these two centers overlap (i.e., $G\ C_n = M\ C_n$), as shown
in Fig. 12.1d.

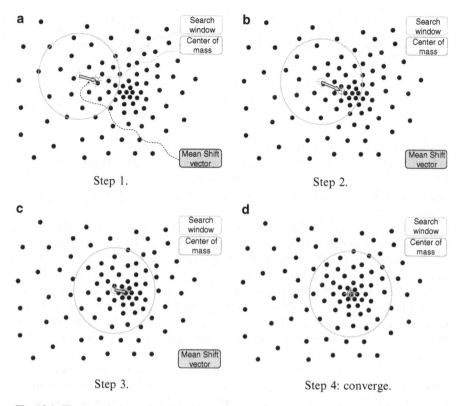

Fig. 12.1 The steps in mean shift algorithm using mass as an example (**a**) Step 1. (**b**) Step 2. (**c**) Step 3. (**d**) Step 4: converge

In image segmentation, each pixel is dealt with as a data point. The coordinate of the pixel is initially set as the geometric center of the window. Then a center of pixel intensity is calculated using the pixels within the window. The geometric center of the window will shift to intensify center of the window until these two centers overlap, i.e., the convergence is reached.

12.3   Hybrid Design of Mean Shift Image Segmentation

The pure software implementation of the mean shift algorithm is adapted from the open source code developed by the Robust Image Understanding Lab at Rutgers University and is based on papers (Christoudias et al. 2002; Comaniciu and Meer 2002). Given an image, the segmentation consists of two major steps. In the first step, the original image is filtered using mean shift method. Two parameters are

Table 12.1 Hybrid design of mean shift segmentation on CPU/GPU platforms

Step		Platform
Mean shift filtering		GPU
Region fusion	Stage 1: Connecting regions	CPU
	Stage 2: Applying transitive closure	Part on CPU, part on GPU
	Stage 3: Pruning spurious regions	Part on CPU, part on GPU

needed, the spatial bandwidth (h_s) and the range bandwidth (h_r). The h_s defines the radius of the window used in mean shift. The filtered image consists of numerous regions. In the second step, i.e., region fusion, three stages are carried out in sequence, including connecting regions, applying transitive closure, and pruning spurious regions.

In our experiments, it is found that the first step, i.e., mean shift filtering, takes significantly longer time than the second step when the image size is smaller than 2,048 ×2,048, therefore the mean shift filtering step is executed on GPU. In the three stages in region fusion, the stage of connecting regions is very short. Therefore it is implemented on CPU. The computation times for both the stage of applying transitive closure and the stage of pruning spurious regions significantly increase when the image size is bigger than 2,048 ×2,048. In these two stages, the parts with good parallelism are implemented on GPU and the others are executed on CPU. The detailed distribution of computation is shown in Table 12.1.

Given a pixel P_a in the source image and a search window with radius h_s, the mean shift process is to repeatedly calculate the mean shift vector $\overrightarrow{Mh}$ (shown in Fig. 12.1) until the squared magnitude of $\overrightarrow{Mh}$, i.e., $\| \overrightarrow{Mh} \|^2$, is less than a threshold ε. The number of elements in vector $\overrightarrow{Mh}$ can be 5 if dealing with color images, i.e., the change of x coordinate (Δx), the change of y coordinate (Δy), and the changes of intensities in RGB ($\Delta R, \Delta G, \Delta B$). For grayscale image, only three elements are needed for $\overrightarrow{Mh}$, i.e., $\Delta x, \Delta y$, and ΔI. If we use S to denote the window centered at P_a, and use s to denote a pixel within the window, Δx can be calculated as (12.1), in which I_s, x_s, and w_s are the intensity, the x coordinate, and the weight of pixel P_s, respectively. The weight of each pixel in an image is specified by a predefined weight map.

$$\Delta x = \sum_{s \in S} K\left(\frac{I_s - I_a}{h_r}\right) w_s x_s, \text{ where } K(x) = \begin{cases} 1 & if \; \| x \| < 1 \\ 0 & if \; \| x \| \geq 1 \end{cases} \tag{12.1}$$

Δy and ΔI are computed in a similar way. A new center of the window P_b is calculated as $\overrightarrow{P_b} \leftarrow \overrightarrow{P_a} + \overrightarrow{Mh}$. Then the mean shift vector to P_b is calculated until the center of the window no longer shifts.

The above computation in the mean shift filtering on a pixel is implemented in a GPU kernel function. When the kernel is launched, each pixel in the source image is handled by a GPU thread. If the size of image is $m \times n$, $m \times n$ threads are created and scheduled to execute on hundreds of processing cores on a GPU device.

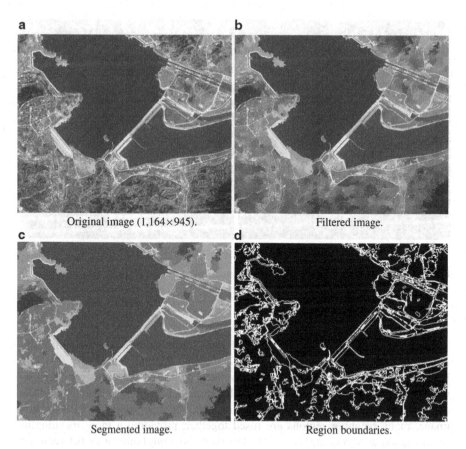

Original image (1,164×945).

Filtered image.

Segmented image.

Region boundaries.

Fig. 12.2 Image segmentation on an airborne image using mean shift (**a**) Original image (1,164 ×945). (**b**) Filtered image. (**c**) Segmented image. (**d**) Region boundaries

12.4   Experiments and Results

The platform is a hybrid workstation consisting of an Intel Core i7-930 quad-core CPU, a Tesla C2075 Fermi GPU (NVIDIA Corporation 2009), and a Tesla K20 Kepler GPU (NVIDIA Corporation 2012). The system has a main memory of 18 GB and runs Ubuntu 10.04.2. The GPU implementation of the mean shift filtering process contains only one GPU kernel function, which is responsible for generating one pixel in the filtered image. Therefore the number of GPU threads is same to the number of pixels in the original image. These threads are grouped into 1-dimensional thread blocks, each of which contains 128 threads. Due to the improved memory hierarchy on Fermi/Kepler GPU architectures, typical performance optimization techniques, such as memory coalescing and memory prefetching (Stratton et al. 2012), are not implemented in the mean shift filtering.

We first applied two versions of the mean shift segmentation algorithm on the image shown in Fig. 12.2a, which is an airborne image of a water dam.

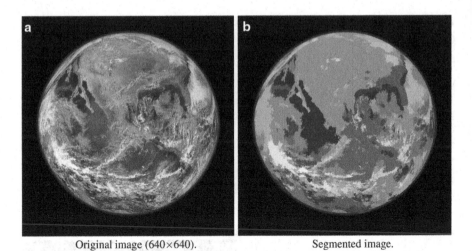

Original image (640×640).   Segmented image.

Fig. 12.3 Image segmentation on a spaceborne image using mean shift (**a**) Original image (640 ×640). (**b**) Segmented image

These two versions include the pure software implementation and Fermi GPU implementation. Figure 12.2b shows the image after applying mean shift filtering. It can be seen that the filtered image has been segmented into numerous regions with unnecessary details. For example, different parts of a hill may belong to different regions. After applying the region fusion step, which is carried on CPU for this image, many neighbor regions are fused together, providing a more meaningful segmentation as shown in Fig. 12.2c. The corresponding boundaries between the regions are shown in Fig. 12.2d. For this airborne image, the software implementation of the mean shift filtering process takes 176.24 s, while the Fermi GPU implementation takes only 11.44 s, achieving a 15 × speedup.

In order to further demonstrate the benefit of the hybrid implementation, we tested three implementations of mean shift filtering on spaceborne globe images of various resolutions. The globe image with resolution 640 × 640 and its segmentation are shown in Fig. 12.3. The clouds are difficult to be distinguished from the oceans and the lands in grayscale images. However, as shown in Fig. 12.3b, mean shift algorithm is capable of providing a meaningful segmentation by separating lands, oceans, and clouds. The performance of mean shift filtering is shown in Table 12.2. It can be found that the performance improvement climbs as the size of image increases until hitting a plateau around 20 × for Fermi GPU and 30 × for Kepler GPU, respectively. When the image resolution increases, more GPU thread blocks are available to be scheduled to stream multiprocessors, resulting in a higher GPU occupancy and a better performance. Once the occupancy is maximized, adding more thread blocks cannot further increase the processing throughput.

It is noticed that the time spent on the region fusion step is negligible compared with the mean shift filtering step when the image size is smaller than 2,048 × 2,048.

Table 12.2 Performance improvement on mean shifter filtering

Image size	Processing time (s)				Image size	Processing time (s)			
	CPU	Fermi	Kepler	Speedup		CPU	Fermi	Kepler	Speedup
240 ×240	4.67	0.42	0.36	11.12 / 12.97	1,024 ×1,024	86.70	5.35	3.44	16.21 / 25.20
320 ×320	8.25	0.62	0.52	13.31 / 15.87	2,048 ×2,048	354.91	18.01	13.37	19.71 / 26.55
500 ×500	19.06	1.37	0.98	13.91 / 19.45	4,096 ×4,096	1321.76	61.59	40.90	21.46 / 32.32
640 ×640	31.26	2.19	1.51	14.27 / 20.70	8,000 ×8,000	4046.15	202.5	139.05	19.98 / 29.10

However, the complexity of the region fusion increases exponentially when the image becomes larger, i.e., $>2,048 \times 2,048$ as shown in Table 12.3. The degree of parallelism in the region fusion step is much lower than the degree of parallelism in the filtering step. In the hybrid implementation on CPU/GPU platform, part of both Stage 2 and Stage 3 is implemented on Kepler GPU. For Stage 2, the hybrid implementation is still able to achieve more than 2 folds of speedup. The speedup on Stage 3 is very marginal.

One other very popular image segmentation algorithm is k-means, which is extremely faster compared with mean shift algorithm. The k-means algorithm is an iterative technique that is used to partition an image into k clusters. The basic algorithm is as follows.

1. Pick k cluster centers, either randomly or based on some heuristic;
2. Assign each pixel in the image to the cluster that minimizes the distance between the pixel and the cluster center;
3. Re-compute the cluster centers by averaging all of the pixels in the cluster;
4. Repeat steps 2 and 3 until convergence is reached (e.g., no pixels change clusters).

The quality of the solution depends on the initial set of clusters and the value of k. The most common form of the algorithm uses an iterative refinement heuristic known as Lloyd's algorithm (Lloyd 1982). Lloyd's algorithm starts by partitioning the input points into k initial sets, either at random or using some heuristic data. It then calculates the mean point, or centroid, of each set. It constructs a new partition by associating each point with the closest centroid. Then the centroids are recalculated for the new clusters, and algorithm repeated by alternate application of these two steps until convergence, which is obtained when the points no longer switch clusters (or alternatively centroids are no longer changed). We implemented the Lloyd's algorithm and applied on the globe image as the results shown in Fig. 12.4. It can be seen that k-means algorithm does not produce quite meaningful segmentation when dealing with complex remote sensing images.

12.5   Conclusions

Mean shift is a robust and non-parametric image segmentation algorithm that is capable of generating better results than other algorithms such as k-means. However, it is a computationally intensive process taking hours to deal with large images. Thanks to the intrinsic parallelism of mean shift algorithm, GPU can be leveraged to accelerate its performance. In this work, the mean shift filtering step is implemented on GPU. The experimental results on a Tesla K20 GPU demonstrate a $30 \times$ speedup compared with an Intel i7-930 CPU for the mean shift filtering step. The hybrid implementation of the region fusion step, however, achieves a much smaller speedup due to a lower degree of parallelism in this step.

Table 12.3 Performance improvement on the mean shifter segmentation algorithm

Image size	CPU implementation time (s)					Image size	CPU/GPU (Kepler) implementation time (s)					Speedup
	Filtering	Region fusion					Filtering	Region fusion				
		Stage 1	Stage 2	Stage 3	Total			Stage 1	Stage 2	Stage 3	Total	
240×240	4.67	0.01	0.01	0.02	4.71	240×240	0.36	0.01	0.01	0.02	0.40	11.78
320×320	8.25	0.01	0.03	0.03	8.32	320×320	0.52	0.01	0.03	0.02	0.58	14.34
500×500	19.06	0.01	0.12	0.11	19.30	500×500	0.98	0.01	0.10	0.10	1.19	16.22
640×640	31.26	0.03	0.33	0.18	31.80	640×640	1.51	0.03	0.19	0.13	1.86	17.10
1,024×1,024	86.70	0.07	0.83	0.67	88.27	1,024×1,024	3.44	0.07	0.40	0.58	4.49	19.66
2,048×2,048	354.91	0.27	18.84	10.41	384.43	2,048×2,048	13.37	0.27	9.49	9.32	32.45	11.85
4,096×4,096	1,321.76	1.26	853.35	576.80	2,753.17	4,096×4,096	40.90	1.26	335.95	517.58	895.69	3.07
8,000×8,000	4,046.15	4.80	12,725.09	8,391.56	25,167.60	8,000×8,000	139.05	4.80	5,385.40	7,730.77	13,260.02	1.90

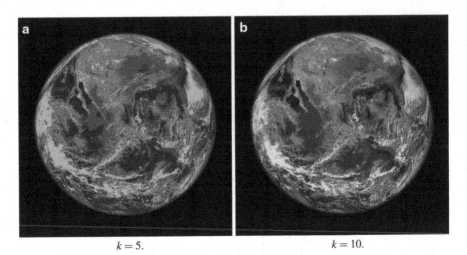

$k = 5.$ $k = 10.$

Fig. 12.4 Segmentation using k-means (**a**) $k=5$. (**b**) $k=10$

Acknowledgements Both Fermi GPU and Kepler GPU used in this work were donated by Nvidia Corporation. The authors would like to thank Dr. John Gauch for the discussion and sharing the source code of k-means algorithm.

References

Cheng Y (1995) Mean shift, mode seeking, and clustering. IEEE Trans Pattern Anal Machine Intell 17(8):790–799

Christoudias CM, Georgescu B, Meer P (2002) Synergism in low level vision. In: Proc. 16th International Conference on Pattern Recognition (ICPR'02), pp 150–155

Comaniciu D, Meer P (2002) Mean shift: A robust approach towards feature space analysis. IEEE Trans Pattern Anal Machine Intell 24(5):603–619

Fukunaga K, Hostetler LD (1975) The estimation of the gradient of a density function, with applications in pattern recognition. IEEE Trans Inform Theory 21(1):32–40

Kanungo T, Mount DM, Netanyahu NS, Piatko CD, Silverman R, Wu AY (2002) An efficient k-means clustering algorithm: Analysis and implementation. IEEE Trans Pattern Anal Machine Intell 24(5):881–892

Lloyd SP (1982) Least squares quantization in PCM. IEEE Trans Inform Theory 28(2):129–137

NVIDIA Corporation (2009) NVIDIA's next generation CUDA compute architecture: Fermi. White paper V1.1, available online on http://www.nvidia.com

NVIDIA Corporation (2012) NVIDIA's next generation CUDA compute architecture: Kepler gk110. White paper V1.0, available online on http://www.nvidia.com

Shapiro LG, Stockman GC (2001) Computer Vision. Prentice Hall, Upper Saddle River, New Jersey

Stratton J, Anssari N, Rodrigues C, Sung IJ, Obeid N, Chang L, Liu GD, Hwu W (2012) Optimization and architecture effects on GPU computing workload performance. In: Proc. 2012 Innovative Parallel Computing: Foundations & Applications of GPU, Manycore, and Heterogeneous Systems (InPar2012), pp 1–10

Part V
Multi-core Technology for Geospatial Services

Chapter 13
Simulation and Analysis of Cluster-Based Caching Replacement Based on Temporal and Spatial Locality of Tiles Access

Rui Li, Xinxing Wang, Jingjing Wang, and Huayi Wu

Abstract Cluster-based caching systems can accelerate users' access to large-scale network services. One of the difficult issues for realizing a cluster-based caching system, however, is how to configure numerous parameters, to make cluster-based caching servers cooperate with each other, to share cached data, and thus obtain optimal performance from the system. This paper analyzes tile access characteristics in networked geographic information systems and simulates cluster-based caching system through a trace-driven experiment based on the log files from the digital earth. Using a collaborative approach between cluster-based caching servers, each parameter in a cluster-based caching system is quantitatively analyzed to obtain a global optimal parameter combination. This research provides a reference for the realization of a cluster-based caching system in networked geographic information applications, to improve the quality of service in networked geographic information systems.

Keywords Cluster • Cache • Replacement • Network GIS

13.1 Introduction

Networked geographic information systems (NGISs) are increasingly popular, with huge numbers of user visits. Scalable network services are needed to meet the increasing quantity of service demands being placed on NGISs. An efficient caching strategy can greatly shorten the response time for users' access requests, to

R. Li (✉) • X. Wang • J. Wang • H. Wu
State Key Laboratory of Information Engineering in Surveying, Mapping and Remote
Sensing, Wuhan University, Wuhan, Hubei 430079, P.R. China
e-mail: RuiLi@whu.edu.cn

X. Shi et al. (eds.), *Modern Accelerator Technologies for Geographic Information Science*, 169
DOI 10.1007/978-1-4614-8745-6_13, © Springer Science+Business Media New York 2013

accelerate the extraction of spatial data (Li et al. 2012). A cache replacement algorithm is the core for caching, and directly affects the utilization of the cache memory system and the efficiency of the cache strategy. A cluster-based framework can overcome insufficient processing ability and limited I/O bandwidth of a single server (Chism and Enos 2006; Barroso et al. 2003). Such an algorithm can improve the reliability and availability of a NGIS service. Therefore, by applying cluster technology to a cache system and the deployment of cached data, better availability and scalability for cache systems are possible, providing better NGISs services in spite of many concurrent requests for spatial data.

Cluster-based cache technology has the following advantages: First, caching can accelerate the data extraction process. A cluster-based cache system caches the data requested by a large number of users, reducing not only the read request times for data storage servers, but also the traffic load. Second, the cluster-based cache servers can cooperate with each other and the localized data in the GIS application server, thereby further saving I/O bandwidth for the data storage servers. Finally, since the cluster-based caches are in the GIS application servers, the cached data can be shared by a large-scale user population (Li et al. 2005) to reduce the response time when users are roaming and to improve the interactivity and quality of service (Wu et al. 2011; Yang et al. 2005a). However, one of the challenging issues is how to configure many parameters in a cluster-based caching system. A good configure can make cluster-based servers cooperate with each other and share cached data, thereby obtaining the optimal performance of the cluster-based caching system. Google Earth, NASA's World Wind, and Microsoft Virtual Earth all use a high performance cluster-based caching system as the system access entrance to achieve good service performance. However, the technological details of cluster-based caching systems have not been published.

Access requests for tiles have temporal and spatial locality (Podlipnig 2003; Baentsch et al. 1997). For example, when a user is interested in a region for a sustained period of time, he will access the spatial data for this region repeatedly. This reflects the temporal locality characteristic of tile access. When a user is accessing the spatial data of a region, the user will also be interested in the spatial data immediately around that region. The accessed area will therefore expand to the adjacent area of the current region. This also reflects the spatial locality characteristic when accessing tiles. However, cache size is limited. When it is filled by the tiles that have been visited and then saved, the latest visited tile cannot be cached. Thus it must delete the data with the "lower value for caching" according to a cache replacement strategy and cache the newly defined hotspot data.

If the cache replacement algorithm considers the behavior of users, analyzes the correlation between the current request and the previous requests, and leaves tiles with higher correlation for access in the cache, the cache hit rate and utilization of the cache will be improved, while replacing tiles that are not likely to be accessed in the near future. Many classic replacement algorithms can do it well, such as Least Frequently Used (LFU) (Dan and Towsley 1990) replacement algorithm, Least Recently Used (LRU) replacement algorithm, First in First out (FIFO) replacement algorithm (Lee et al. 1999), the replacement algorithm based on the specific access content, or transmission time cost, such as size, the weights of the last access time

(Cao and Irani 1997), etc. Google and NASA have used the LRU replacement algorithm in their caching system (O'Neil 1993; Boulos 2005; Bell et al. 2007). All these algorithms did not involve multi-server cluster-based caching or a collaborative approach for a cluster-based caching system. A cache replacement algorithm is the soul of a cluster-based caching system (Yang et al. 2005b; Barish and Obraczke 2000). The configuration of the parameters in a cluster-based caching system is critical for whole system performance. Network simulation efficiently obtains the optimal global parameters for cluster-based caching and secures valuable experimental results to focus future research.

This paper discusses the cache replacement algorithm and its relation to the temporal and spatial locality characteristics of tile access. Using a trace-driven experiment based on the server log files from the digital earth system developed by the authors, this paper simulates cluster-based caching and a collaborative approach between cluster-based caching servers. By quantitatively analyzing each parameter of the cluster-based caching system, it obtains the global optimal parameter combination, and provides a measurable reference for deploying cluster-based servers in NGISs applications meeting the quality of service needs for NGISs.

13.2 Caching Replacement and the Temporal and Spatial Locality of Tile Access

Localized accessed hotspots exist since 80 % of requests target only 20 % of the data (Wang 2009). The locality characteristics in tile access are featured as temporal locality and spatial locality. Many caching replacement strategies consider the locality characteristics to yield better performance of services.

13.2.1 The Temporal Locality of Tile Access

Temporal locality implies that recently accessed tile has a higher probability of being accessed again in the future (Wang 2009). The interval between current time and the user last access time to a tile is a representation of temporal locality. LRU is a typical algorithm based on temporal locality. It considers that access probability is inversely proportional to the interval between the current time and the user last access time. The probability can be stated as formula (13.1). p is the access probability, Δt stands for the interval time, and $\leftarrow$ is a proportional symbol.

$$p \leftarrow 1 / \Delta t, \Delta t = current\ System\ Time - last\ Access\ Time. \qquad (13.1)$$

LRU reflects the short-term popularity of the accesses to tiles (Lee et al. 1999). When ranking the access probability of a tile according to LRU by descending order, the tiles which have been accessed recently are placed on the top rank and the tiles which have been accessed at an earlier time are places on the lower rank. The tiles are ranked by the last access time.

The well-known First In First Out (FIFO) algorithm where tile access probability is inversely proportional to the interval between current time and the first access time. It can be expressed as:

$$p \leftarrow 1 / \Delta t, \Delta t = current \; System \; Time - first \; Access \; Time. \tag{13.2}$$

The tiles are ranked by the first access time according to FIFO.

13.2.2 The Spatial Locality of Tile Access

Spatial locality of tile access implies that the tiles tend to be accessed at adjacent time if they have a spatially adjacent relationship (Wang 2009). If a tile is accessed at a given time, the tiles around it, including the tile itself, have a higher probability of being accessed again at the next time. The space distance between the currently-accessed tile and the tiles in the storage system is a representation of spatial locality. LFU is the typical algorithm, based on a power law (Wu et al. 2011), using the long-term popularity of tile access. Another tile access probability is proportional to the tile access frequency. The probability can be expressed as:

$$p \leftarrow total \; Access \; Times. \tag{13.3}$$

LFU uses the access times to measure the importance of spatial data. The more access times the more importance of the spatial data. LFU takes into account the hot distribution of user's requests to a certain extent.

13.3 Simulation and Analysis

The aim of the simulation is to obtain the optimal combinations of parameters, to improve the cache hit rate and thus reduce the response time in a cluster-based caching system. First, the user's requests were recorded in detail for high-resolution raster data as log files. Then, these log files were input into the trace-driven simulation experiments.

13.3.1 Cluster-Based Caching Replacement Model

Cluster-based caching technology is a combination technology of cluster technology and caching technology. It unites these cache servers with cluster technology, providing a transparent and accelerated service for users' access. Figure 13.1 is a simple cluster-based caching system model.

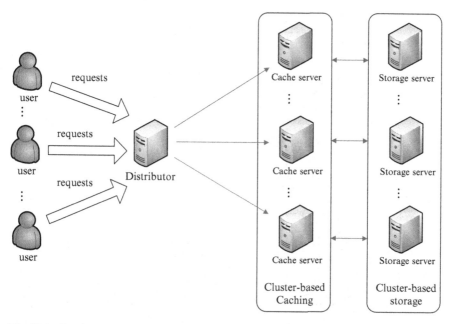

Fig. 13.1 Simple cluster-based caching system model

In this model, users send requests over the Internet. When the requests arrive at the load distributor, the load distributor chooses a server to respond to the requests according to the load balancing strategy.

Based on this model, the collaboration workflow of cluster-base caching system was designed as shown in Fig. 13.2. The load balancing server distributes requests to one of the servers in the cluster-based caching system, according to a load balancing strategy. If a requested tile was hit, namely the tile in the selected server, it is termed a "local hit". If the requested tile was hit on other server of the cluster-based caching system, it is termed a "cluster hit". If it does not hit anywhere in the whole cluster-based caching system, it is termed "cluster missing". For requests with a cluster missing status, we start the cache replacement algorithm if the number of current caching tiles exceeds the replacement threshold value, and replace the tiles with the lower caching value by the tiles with the higher caching value. It maintains a reasonable number of the cached tiles, to improve the performance of the cluster-based caching system.

Figure 13.2 shows the key parameters in cluster-based cooperative caching system include the load balancing algorithm, the neighbor selection algorithm, then cache size, the number of cache servers, the replacement threshold value, and the cache replacement algorithm. There has been a great deal of scientific research about load balancing and neighbor selection algorithms, however, and a detailed analysis of these algorithms is beyond the scope of this paper. Thus, the classic round-robin algorithm is efficient for load balancing was used in the simulation.

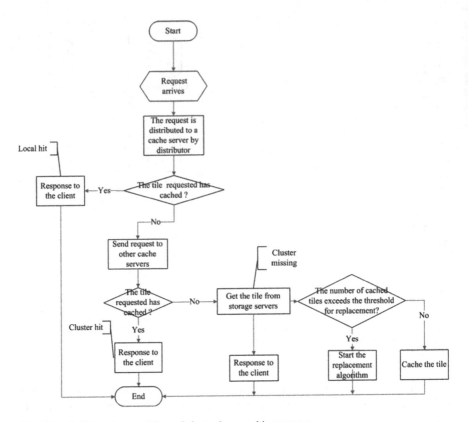

Fig. 13.2 Collaboration workflow of cluster-base caching system

Thus, the simulation only addresses the size of the cache, the number of cache servers, the replacement threshold value and the cache replacement algorithm, and their impact on the cooperation workflow of the cluster-based caching system. The replacement threshold value is the threshold value for the size of current cached tiles when it needs start the cache replacement algorithm.

13.3.2 Simulation Design

The paper describes a simulation of the cluster-based caching system in a networking simulation environment. Twelve Linux servers were connected via a 1,000 Mbps switch to form a fast Ethernet. The number of user access requests arriving on the cluster-based caching system in a unit time is smooth, with no aftereffect, and demonstrate universality. A Poisson distribution was used to describe users' access requests arriving. The process-time of a request on a server is memory-less, described by a negative exponential distribution. The cache replacement algorithms,

Table 13.1 Simulation configurations

Clients	360 clients with 6 fast Ethernet connections, connected by an Ethernet switch with 64 ports. The number of tile requests are 100,000.
Load-balancing server	A Linux Server with Intel Xeon E5620 2.40 GHz, quadruple-core processor, 8GB RAM. The request arrival rate for tile access obeys a Poisson distribution.
Cluster servers	12 Linux servers with Intel Xeon E5620 2.40 GHz, quadruple-core CPU, 8GB RAM, connected via a 1,000 Mbps switch to form a fast Ethernet. The server time for processing of a request obeys a negative exponential distribution.
Spatial data	90 m of global SRTM terrain data, size 128×128, and 30 m of global Landsat7 satellite image data, size 512×512.

such as FIFO, LRU and LFU, are classical algorithms, and can effectively reflect the locality of the access to tiles. Thus, the experiments in this paper use them as a parameter for the cache replacement algorithms and compare their performance. Based on the cluster-based caching collaborative model, a load dispatcher with sufficient processing power is placed at the entrance to the cluster system to prevent forwarding bottlenecks. The cache hit rate gives an indication of the accuracy of the cache policy. In the NGISs, an important objective of a clustered cache system is to reduce the average request response time, making user navigation faster and smoother. Therefore, the simulation takes the cache hit rate and the average request response time as indicators of the performance of a cluster cache system. The simulation configurations are as follows (Table 13.1).

13.3.3 Simulation of Cache Size

There are two parameters that affect the cache hit ratio. One is the cache size and the other is replacement threshold value. To simplify the experiment, this paper will first consider these two parameters on a single-server. The relative size of the cache expresses the size of cache, which is the ratio of the cache size to the total size for the tiles requested.

The replacement threshold value was set to 100 %, starting the cache replacement algorithm when the cache space is used up. Then, the relationship between relative size of the cache and the cache hit rate can be examined. FIFO, LRU, and LFU simulation results are compared, while the relative cache size is increased from 10 to 100 % by step of 5 %.

As described in Fig. 13.3, the hit rate of FIFO is lower than the LRU and LFU hit rates. The cache hit rates for LRU and LFU are almost the same. This is because LRU considers only the temporal locality characteristic of tile access while LFU considers only the spatial locality characteristic. However FIFO does not take into account either temporal or spatial locality, nor considers other characteristics of tile access patterns. LRU and LFU can improve the cache hit rate, if the temporal or

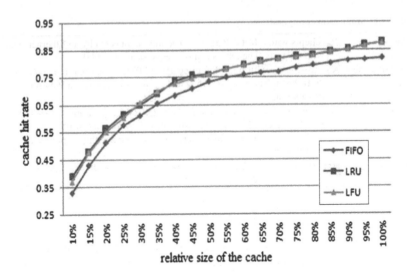

Fig. 13.3 Cache hit rate for different relative sizes of the cache

spatial locality of tile access is considered and the increasing levels of both methods in cache hit rate are similar. When the relative size of the cache increases, the value of the cache hit rate has an upward trend, but the rate of increase gradually slows down.

Figure 13.4 describes the growth rate of the cache hit rate when the relative cache size increases. f(x) stands for the cache hit rate when relative size of the cache is x.

$$\text{Growth rate} = f(x+1) - f(x) / f(x) \tag{13.4}$$

As illustrated in Fig. 13.4, when the relative size of the cache is increases, the growth rate of the hit rate gradually reduces. When the relative size of the cache is smaller (15–30 %), the cache hit rate increases quickly with an increasing cache size. When the relative size of the cache is 35–50 %, the cache hit rate increases slowly, while the growth rate is almost 5 %. When the relative size of the cache is more than 60 %, the growth rate remains steady at almost 2 %. This indicates that enlarging the cache size has a little impact on caching performance when the relative size of the cache is more than 60 %.

13.3.4 Simulation of Replacement Threshold Value

Section 13.3.3 shows that the cluster based server's cache is wasted when the relative cache size is increased to over 60 %. The simulation selects 60 % as the optimal value for the relative cache to study the replacement threshold value parameter.

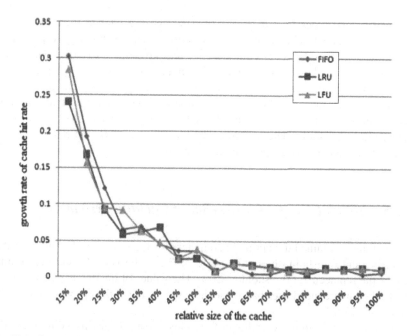

Fig. 13.4 Growth rate of cache hit rates for different relative sizes of the cache

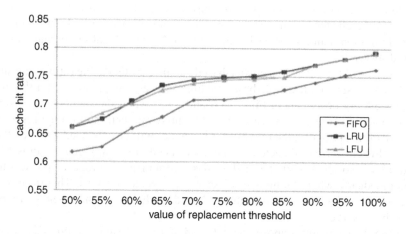

Fig. 13.5 Cache hit rate for the value of replacement threshold

The value of replacement threshold ranges from 50 to 100 %, and the step is 5 %. Three different replacement algorithms were simulated.

Figure 13.5 shows that the cache hit rate increases monotonically when the replacement threshold value increases, the higher the replacement threshold value, the higher the cache hit rate.

However, two types of data, tiles requested and pre-fetched tiles, are usually stored in the cache. Cache prefetching is an active cache technology. It prefetches tiles that are likely to be accessed in advance and saves them to cache. The higher the value of the replacement threshold, the smaller the space left for pre-fetched tiles. Considering the correlation of caching technology and prefetching technology, their impacts on the cache hit rate, and the fact that a cache hit rate is proportional to the value of replacement threshold, 95 % was chosen as the best value for a replacement threshold. The missed cache hit rate caused by the remaining 5 % cache can be made up with a pre-fetching strategy.

13.3.5   Simulation of a Cluster-Based Caching System

The simulation obtained the optimal parameters for the size of cache and the value of replacement threshold, the optimal size of the cache is 60 % and the optimal value of the replacement threshold is 95 %. The values of the two parameters were then used simulate the cluster-based caching system, to get the optimal value of the number of servers.

It can be seen from the workflow of the cluster-based caching system that the cache hit rate includes the local hit rate and cluster hit rate. Using the cache hit rate as an indication of the performance of the cluster-based caching system will make the simulation complicated, because it needs to record both the local hit rate and the cluster hit rate. Therefore, this paper counts the cache miss-rate in this simulation, namely the cache miss-rate on both the local server and cluster-based servers.

The size of the cache in a cluster-based caching system is the ratio of the sum of the cache size of all cluster-based caching servers to the total size for the tiles requested. As described in Figs. 13.6 and 13.7, when the number of cluster-based caching servers increases, the cache miss-rate and the average response time decline gradually; the reliability of services will be enhanced. On the other hand, this cluster-based caching system will need more hardware. Figures 13.6 and 13.7 also show that the cache miss-rate and the average response time become steady when the number of servers is greater than 16. Therefore, it is useless to add servers to the cluster-based caching system when there are already 16 servers.

From the simulations above, we obtained a group of values of optimal parameters: the value of the relative cache space is 60 %, the value of replacement threshold is 95 %, and the number of the cluster-based servers is 16. If cache replacement algorithms follow the characteristics of the user access pattern for tiles, and consider the temporal and spatial locality of tile access, they will yield better performance.

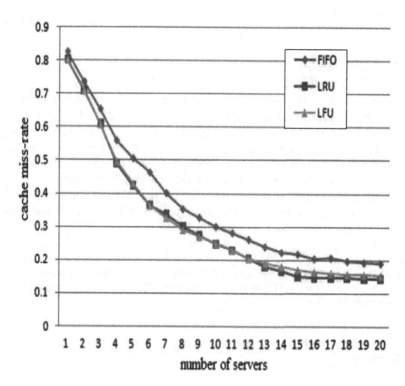

Fig. 13.6 Cache miss-rate for cluster-based servers

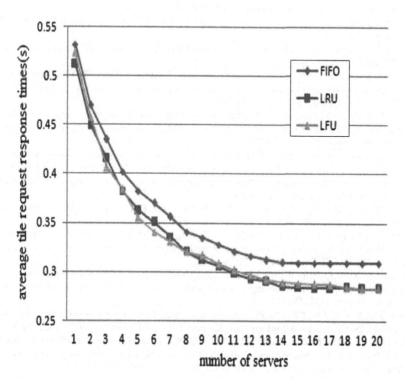

Fig. 13.7 Average tile request response times for cluster-based servers

13.4   Conclusions

Cluster-based cache replacement is a combination of cluster and cache technologies and a cache replacement algorithm. The key to cluster-based caching systems is to configure the many parameters in cluster-based systems, before services are provided, so as to get a higher hit rate and accelerate the responses to tile requests. Simulation results demonstrate that the optimal set of parameters as obtained from the simulation can improve the utilization efficiency of cluster-based systems to yield better performance. In future research, both the temporal and spatial locality of tile access patterns will be considered, toward a better cache replacement algorithm to gradually reduce the request response time.

Acknowledgements This work was supported by the National Natural Science Foundation of China (Grant No. 41071248), Project supported by the Foundation for Innovative Research Groups of the National Natural Science Foundation of China (Grant No. 41021061), and the LIESMARS Special Research Funding.

References

Baentsch M, Baun L, Molter G, Rothkugel S, Sturm P (1997) World wide web caching: the application-level view of the Internet. IEEE.COMM.M 35:170–178

Barish G, Obraczke K (2000) World wide web caching: Trends and techniques. IEEE.COMM.M. 38:178–184

Barroso L A, Dean J, Holzle, U (2003) Web search for a planet: The Google cluster architecture. IEEE micro 3(2):22–28

Bell D G, Kuehnel F, Maxwell C, Kim R, Kasraie K, Gaskins T, Hogan P, Coughlan J (2007) NASA World Wind: Opensource GIS for Mission Operations. In: 22nd Asian Conference on Remote Sensing. IEEE Press, Big Sky, MT, pp 1–9

Boulos M NK (2005) Web GIS in practice III: creating a simple interactive map of England's strategic health Authorities using Google maps API, Google earth KML, and MSN virtual earth map control. INT.J.HE.GEO 4(22)

Cao P, Irani S (1997) Cost-aware WWW proxy caching algorithms. In: Proceedings of the USENIX Symposium on Internet Technologies and Systems. USENIX Press, California, pp 193–206

Chism F, Enos J (2006) Running a Top-500 benchmark on a windows compute cluster server cluster. In: ACM New York, NY, 2006

Dan A, Towsley D (1990) An approximate analysis of the LRU and FIFO buffer replacement schemes. ACM.SIGMETRICS. P.E. R. 18:143–152

Lee D, Choi J, Kim J H, Noh S H, Min S L, Cho Y, Kim C S (1999) On the existence of a spectrum of policies that subsumes the least recently used (LRU) and least frequently used (LFU) policies. ACM.SIGMETRICS. P.E. R. 27:134–143

Li R, Guo R, Xu, Z Q, Feng W (2012) A prefetching model based on access popularity for geospatial data in a cluster-based caching system. International Journal of Geographical Information Science 26 (10):1831–1844

Li H S, Zhu X Y, Li J W, Chen J (2005) Research on Spatial Data Distribution Cache Technology in WebGIS. Geomatics and Information Science of Wuhan University 30(12):1092–1095

O'Neil E J, O'Neil P E, Weikum G (1993) The LRU-K page replacement algorithm for database disk buffering. In: Proceedings of the 1993 ACM SIGMOD international conference on Management of data. ACM New York, NY, USA, pp 297–306

Podlipnig S, BÖSZÖRMENYI L (2003) A survey of web cache replacement strategies. ACM.C.SURV 35:374–398

Wu H Y, Li Z L, Zhang H W, Yang C, Shen S (2011) Monitoring and Evaluating Web Map Service Resources for Optimizing Map Composition over the Internet to Support Decision Making. Computers and Geosciences, 37:485–494

Wang H (2009) Research on distributed load balancing and cache technologies for multimedia networked GIS. Ph.D., Dissertation of Wuhan University

Yang C, Wong D W, Yang R, Kafatos M (2005) Performance-improving Techniques in Web-based GIS. INT.J.GEO.I. 19:319–342

Chapter 14
A High-Concurrency Web Map Tile Service Built with Open-Source Software

Huayi Wu, Xuefeng Guan, Tianming Liu, Lan You, and Zhenqiang Li

Abstract As an Open Geospatial Consortium (OGC) standard, the Web Map Tile Service (WMTS) has been widely used in many fields to fast and conveniently share geospatial information with the public. In practice, however, when client users increase dramatically, the torrent of requests places overwhelming pressure on the web server where WMTS is deployed, causing significant response delay and serious performance degradation. The architecture of high-concurrency WMTS therefore must be extended to automatically scale with the requests of client users. This paper introduces a prototype for such a high-concurrency WMTS, built totally with open-source software, including Nginx, GeoWebCache, and MongoDB. Several experiments were carried out to test the efficiency and scalability of the proposed high-concurrency WMTS using Web-bench. The results illustrate that our WMTS can function well even when enduring more than 30,000 concurrent connections. The request throughput of the proposed high-concurrency WMTS is twice as large as that of traditional WMTS deployed in a single web server.

Keywords WMTS • Web service • High-concurrency • Reverse proxy • NoSQL

14.1 Introduction

To improve the performance of Web Map Service (WMS), OGC released a standard for tile-based web mapping, Web Map Tile Service (WMTS) (Open Geospatial Consortium Inc. 2010). Instead of creating a new image for each request, the WMTS

H. Wu • X. Guan (✉) • T. Liu • L. You • Z. Li
State Key Laboratory of Information Engineering in Surveying, Mapping and Remote Sensing, Wuhan University, 129 Luoyu Road, Wuhan 430079, P.R. China
e-mail: wuhuayi@whu.edu.cn; guanxuefeng@whu.edu.cn

X. Shi et al. (eds.), *Modern Accelerator Technologies for Geographic Information Science*, 183
DOI 10.1007/978-1-4614-8745-6_14, © Springer Science+Business Media New York 2013

returns small pre-generated images to users. WMTS provides an open-source alternative to proprietary web mapping services, such as Google Maps, Microsoft Bing Maps.

A Web Map Tile Service (WMTS) is a fast, convenient, and highly efficient way to share geographic information with the public over the internet. Because of its efficiency, WMTS has been increasingly adopted in many applications, such as public service, navigation, location-based services, and social networking. However, there still exist problems with WMTS scalability in practice, i.e. how WMTS servers handle massive concurrent requests. When client users increase dramatically, the torrent of client requests places overwhelming pressure on the web server where the WMTS is deployed, causing significant response delay and serious performance degradation of the WMTS. Therefore, it is of great necessity to study the architecture for high-concurrency WMTS that can automatically scale with the requests of client users.

This paper fills the gap in the research concerning high concurrency WMTS and implements a prototype system on a high performance cluster (HPC). The high-concurrency WMTS is totally built with open-source software, including MongoDB, GeoWebCache, and Nginx. In this prototype system, MongoDB is used to store massive tile images in the HPC; GeoWebCache is customized to publish WMTS service and process requests from users; Nginx acts as a powerful load balancer to client requests. Several experiments were carried out to evaluate the scalability of the proposed high-concurrency WMTS with Web-bench. The results illustrate that our WMTS can function even when enduring more than 30,000 concurrent connections. The request throughput of this proposed high-concurrency WMTS is twice as large as that of a traditional WMTS deployed in a single web server.

The rest of the paper is arranged as the following. Section 14.2 presents related work on the research of high-concurrency system and web service benchmarks. Section 14.3 explains the architecture of high-concurrency WMTS. Section 14.4 presents the results and discussion of the experiments. Section 14.5 closes the paper with our conclusions.

14.2   Related Work

14.2.1   High-Concurrency Web Services

High-concurrency systems have been an active research field for many years. Presently, there is large body of work in this field, but the following literature is representative of existing research directions and solutions.

Sharifian et al. (2008) built and validated a prototype system on cluster web servers and proposed a load balancing algorithm. The Intelligent Queue-based Request Dispatcher (IQRD) aims to achieve better load balancing with the help of request classification, performance isolation, and dynamic remaining capacity estimation mechanisms.

Yang et al. (2006) carried out research on modeling and simulation of performance analysis of a cluster-based Web server consisting of five real servers. Three ways of load balancing are introduced, including network address translation, IP tunneling, and direct routing. After evaluation and tuning, the maximum process capability of this system is identified and a conclusion is achieved that increasing the number of real servers can be used to tackle the performance bottleneck.

Gautam (2002) investigated the location effects of web proxy servers. His research addressed how to determine the optimal number and locations of proxy servers in a network to minimize costs associated with delay, throughput, and demand constraints. An algorithm called DEJAVU is proposed to solve this optimization problem. The DEJAVU algorithm can take less than a minute to achieve nearly same optimal results compared with genetic algorithms.

Faour et al. (2006) presented a cluster-based web server named Weblins with a fully distributed architecture. Weblins implements a content-aware request distribution policy. Simulation results show that the policy used by Weblins is more suitable for cluster based Web servers in comparison with a pure content-aware strategy or pure cooperative caching strategy.

Chiang et al. (2008) implemented an efficient and scalable web cluster, LVS-CAD/FC (i.e. LVS with Content-Aware Dispatching and File Caching). In LVS-CAD/FC, a content-aware web switch based on TCP Rebuilding is implemented to examine and distribute the HTTP requests to web servers. Besides, a file-based web cache stores a small set of the most frequently accessed web files in server RAM to reduce disk I/O. Experimental results show that LVS-CAD/FC is efficient and scales well. Besides, LVS-CAD/FC with the proposed policies can achieve 66.89 % better performance than the Linux Virtual Server with a content-blind web switch.

These above-mentioned research works can be divided into two categories, load-balance strategies and performance tuning. Basically these research works have focused on the general web server and these solutions cannot deal with the dynamics and huge data throughput of geospatial web services. However, all these research works provide a valuable foundation for high-concurrency WMTS.

14.2.2   Benchmark of Web Services

Dependencies are very common between requests in a web service session, i.e. some later requests depend on the responses of earlier requests in one session. Based on request dependency, the benchmark systems of web services can be divided into request-based systems and session-based systems. In the request-based systems, same or random requests are sent to the target web services in a batch mode. Compared with request-based systems, the core of session-based systems is to model real users and mimic each request session. Thus, the workload generation for session-based systems must address the issue of handling request dependencies.

Numerous benchmark tools have been developed, e.g. S-Client (Banga and Druschel 1999), httperf (Mosberger and Jin 1998), SURGE (Ferrari 1984), WAGON

(Liu et al. 2001), and SWAT (Krishnamurthy et al. 2006). These general purpose workload generators support common workload characteristics, which can be tuned for many different workloads of interest. These characteristics contain queuing delays, idle time, session length, think time, etc. S-Client, httperf, and SWAT use the single-threaded, event-based design, whereas the other tools listed employ the multi-threaded paradigm.

However, the access patterns for geospatial web services are much different from other web services. For example, the web application, Microsoft HotMap (Fisher 2007), represents heat areas in the requests to Bing Maps service. Illustrated by HotMap, tile popularity tends to follow human population. Therefore, new benchmark system for geospatial web services must be developed to address these special access patterns.

14.3   The Architecture of High-Concurrency WMTS

This paper introduces the architecture of a proposed high-concurrency WMTS and implements a prototype system on a high performance cluster (HPC). This high-concurrency WMTS integrates several cutting-edge information technologies, such as reverse proxy, NoSQL database, and server benchmark. It is built totally with open-source software, including MongoDB, GeoWebCache, and Nginx.

14.3.1   Web Map Tile Service

The OGC has specified a suite of web service standards supporting geospatial interoperability, e.g. WMS, Web Feature Service (WFS) and Web Coverage Service (WCS), etc. The WMS specification supports the publishing of cartographic maps on the internet in an interoperable manner.

Currently WMS is widely accepted as an open standard for map visualization and implemented by the majority of GIS software vendors. It standardizes the way in which web clients request maps. Clients can request maps from a WMS provider by specifying map layers and providing parameters such as the size of the returned map and the spatial reference system. When the WMS server receives the client requests, it will generate each requested map image on the fly and realize an almost "instant" zoom and pan for users. However, this flexibility comes at a price. Due to inefficient on-the-fly generation, WMS cannot scale well with client requests.

To improve the performance of WMS, three types of web mapping services were proposed, including WMS-C, Tile Map Service, and WMTS. The first two services were proposed by Open Source Geospatial Foundation (OSGeo), and the last one was proposed by OGC.

The goal of a WMS-C is to find a way to optimize the delivery of map imagery across the Internet. WMS-C defines a constrained profile of OGC WMS that permits servers to optimize their image generation, and allows tiles to be cached at

Fig. 14.1 The illustration of tiling model in the WMTS specification

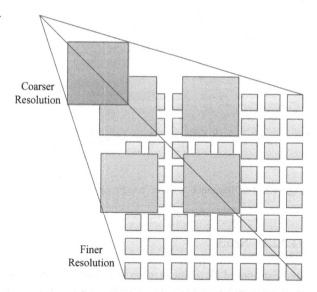

Coarser
Resolution

Finer
Resolution

intermediate points. A WMS-C service is likely to only deliver images for bounding boxes aligned to a given rectangular origin and grid, and only at some fixed scale levels. A WMS-C server should be free to return an exception or a redirect, if it receives a WMS request that is not WMS-C compliant, or involves a bounding box that does not correspond to a single tile in the cache.

A Tile Map Service (TMS) provides access to cartographic maps by predefined cached images, not by direct access to the data itself. The TMS renders spatial data into cartographic tiles at fixed scales. These predefined tiles are provided via a "REST" interface, starting with a root resource describing available layers, then map resources with a set of scales, then scales holding sets of tiles.

WMTS is an evolution of OSGeo's Tile Map Service. WMTS uses a tiling model to describe the predefined images. A tiling model divides the space into a fixed tile matrix, illustrated by Fig. 14.1. The WMTS server can simply return the appropriate pre-generated image (e.g., PNG or JPEG) files to client users. In addition, WMTS supports multiple architectural patterns—KVP, REST and SOAP.

14.3.2   The Architecture of High-Concurrency WMTS

The architecture of high-concurrency WMTS can be segmented into three tiers of services. The three tiers, from top to bottom, are load balancer, WMTS application, distributed tile database. In this architecture, Nginx acts as a powerful load balancer for client requests; GeoWebCache is customized to publish WMTS service and to process user requests; while MongoDB is used to store massive tile images in the HPC.

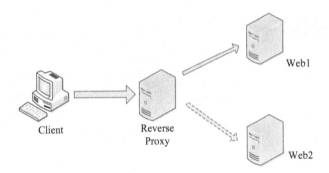

Fig. 14.2 The illustration of the reverse proxy

14.3.2.1 Nginx

In the proposed high-concurrency WMTS architecture, a reverse proxy is used as a load balancer to dispatch the incoming tile requests. A reverse proxy is placed in front of a Web server or cluster servers to handle incoming requests. Reverse proxy parses the request parameters, route incoming requests to one web server, retrieves result information from the web server and then forwards it to the user. A reverse proxy can also be deployed to handle SSL acceleration, intelligent compression, and caching. Being different from a traditional forward proxy located near clients, the reverse proxy acts as an intermediary for its associated servers. The function of one reverse proxy can be illustrated by Fig. 14.2.

Many web servers can be deployed as a reverse proxy, such as Nginx (Reese 2008), HAProxy, and Apache, etc. Among them, Nginx is the second most popular open source web server on the Internet. Its functionality includes HTTP server, HTTP and mail reverse proxy, caching, load balancing, compression, request throttling, connection multiplexing and reuse, SSL offload and HTTP media streaming. It can process 100,000+ concurrent connections per server. The core configuration of the Nginx in the proposed high concurrency WMTS architecture is listed as follows when acting as reverse proxy. This configuration means all requests to the front-end node (192.168.0.151) in the port 80 are redirected to the port 8080 of upstream web servers (from 192.168.0.152 to 192.168.0.15x).

```
upstream web_servers {
    server 192.168.0.152:8080;
    ...
    server 192.168.0.15x:8080;
}
server {
    listen 192.168.0.151:80;
    server_name proxy;
    access_log /var/log/nginx/proxy.access.log;
```

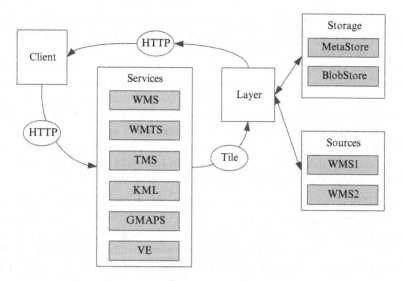

Fig. 14.3 The workflow of processing client request in GeoWebCache

```
error_log /var/log/nginx/proxy.error.log;
location / {
  proxy_pass http://web_servers;
  }
}
```

14.3.2.2   GeoWebCache

In the proposed high-concurrency WMTS architecture, GeoWebCache provide an implementation of the required WTMS service. GeoWebCache is a Java web application used to cache map tiles coming from a variety of sources, such as WMS, or pre-generated tile storage. It implements various service interfaces (such as WMS-C, WMTS, TMS, Google Maps KML, Virtual Earth) in order to accelerate and optimize map image delivery. It can run as a standalone application or be integrated with web map server, e.g. GeoServer. Figure 14.3 shows the process workflow of client request in GeoWebCache.

14.3.2.3   MongoDB

MongoDB, one of the most popular NoSQL databases, is selected to store massive tile images. As opposed to the widely-used relational database management system (RDBMS), NoSQL database is a class of database management systems identified by the following features.

1. *Schema-free*. In a NoSQL database, data can be stored without defining a rigid
   database schema. The data stored are normally self-descriptive. This schema-free
   feature provides immense flexibility for users.
2. *Auto-sharding*. NoSQL database systems usually employ a distributed architecture. The database automatically spreads data across servers without user partici-
   pation. In this way, the system can easily scale out by adding more servers, and
   failure of one server can be tolerated. This type of database typically scales hori-
   zontally and is used for managing massive amounts of data.
3. *Distributed query support*. NoSQL database systems possess the ability to perform complex data queries. They can retain full query expressive power even
   when distributed across hundreds or thousands of servers.

MongoDB is one of the most widely used NoSQL databases available in the current marketplace. Besides the common features of NoSQL databases, MongoDB
stores data in the format of JSON-style document. It supports full indexing on any
attribute of stored documents. One WMTS tile record is illustrated as follows. The
two entries, level and location, provides the position information of one tile in the
tiling pyramid. A 2D spatial index is ensured in the location entry. The tile data is
stored as a binary object in the content entry.

```
mongos> db.EPSG_900913_ 8.findOne()
{
"_id" : ObjectId("4f0008f8db7ecff1831b15f3"),
"name" : 10313,
"level" : 8,
"location" : {
                "col" : 56,
                "row" : 174
            },
"hitTimes" : 0,
"CacheTime" : null,
"content" : BinData(0,"/9j/4AAQSkZA****WxgULUpOx//Z")
}
```

14.3.2.4 The Distributed High-Concurrency WMTS Architecture

The proposed distributed high-concurrency WMTS architecture is built on a standard SMP (Symmetric Multiprocessor) cluster, illustrated in Fig. 14.4. In a SMP
cluster, each node is equipped with two or four symmetric processors. Each processor is multicore-enabled. Thus, there are two levels of parallel computing resources
available for one SMP cluster: inter-node and inner-node parallelism. The whole
distributed high-concurrency WMTS architecture is illustrated in Fig. 14.5.

A prototype was built for the high-concurrency WMTS following the proposed
architecture. The prototype system is constructed on a 16-node virtual Linux cluster
running CentOS 6.2. Each node has two CPUs, 3GB memory, and 100GB hard disk.

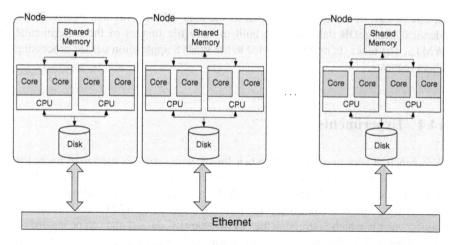

Fig. 14.4 The illustration of a standard SMP cluster

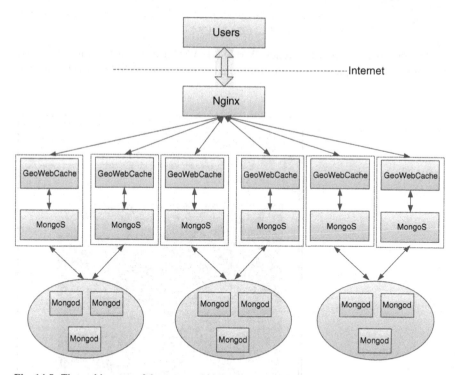

Fig. 14.5 The architecture of the proposed high-concurrency WMTS

In this cluster, one node is configured as the load balancer running Nginx. Three identical MongoDB databases are built to store tile images of the experimental WMTS. The other six nodes are added to the WMTS application tier for processing client requests.

14.4 Experiments

Web-bench, a very powerful tool to benchmark web server, is used to evaluate the request throughput of the proposed high-concurrency WMTS. It will simulate hundreds of thousands of clients to issue WMTS requests. The URL and result of one WMTS tile request are listed in Fig. 14.6. The following WMTS URL means to request one tile which is stored in the "EPSG:900913" dataset and can be located by "EPSG:900913:3", 56 and 174 (i.e. level, row and column respectively).

```
http://192.168.0.151/geospeed/service/wmts?request=GetT
    ile&version=1.0.0&layer=Shanghai&style=default&format
    =image/gif&TileMatrixSet=EPSG:900913&TileMatrix=EPSG:
    900913:3&TileRow=56&TileCol=174
```

The proposed high-concurrency WMTS(H-WMTS in the following tables and figures) was compared with a traditionally deployed WMTS(WMTS in the

Fig. 14.6 The URL and result of one WMTS tile request

Table 14.1 Request throughput (pages/min) from concurrent connections

Conn.	2,000	4,000	6,000	8,000	10,000	12,000	14,000	16,000
WMTS	195,814	205,202	208,376	209,110	204,090	202,072	187,708	190,506
H-WMTS	50,130	269,708	281,002	332,124	291,974	382,162	363,644	400,108
Conn.	18,000	20,000	22,000	24,000	25,000	26,000	28,000	30,000
WMTS	194,908	188,650	198,416	200,234	198,368	204,206	204,950	127,820
H-WMTS	424,956	414,922	431,064	422,440	435,130	418,470	403,668	245,262

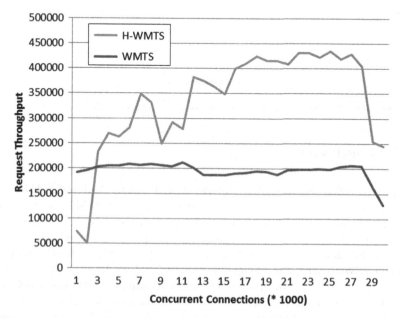

Fig. 14.7 The request throughput comparison between two types of WMTS

following tables and figures), in which client requests are directly sent to one GeoWebCache and tile images are stored in a MongoDB database on three computer nodes. The Web-bench commands are listed as following (-c means the number of concurrent connections; -t means the running time of continuous requests; and wmts_url is the request URL for one specific tile). Here, the continuous request time is defined to 30 s.

```
webbench -c n -t 30 wmts_url
```

A comparison of request throughput (requests per minute) for high-concurrency WMTS and traditionally deployed WMTS is shown in Table 14.1 as well as in Figs. 14.7 and 14.8. Figure 14.7 describes that, when the concurrent client connections is about 30,000, the high-concurrency WMTS can handle nearly 420,000 request per minute while the traditionally deployed WMTS only can process 200,000 request in a minute.

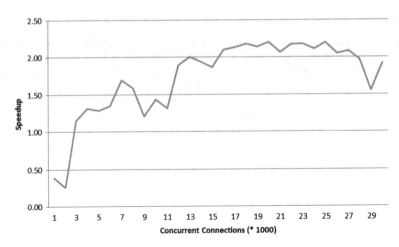

Fig. 14.8 The speedup of request throughput between two types of WMTS

14.5 Conclusion

In order to tackle the scalability problem of traditionally deployed WMTS, this paper proposes a novel architecture of high-concurrency WMTS with several cutting-edge information technologies, such as reverse proxy, NoSQL database, and server benchmark. We also implemented a prototype system to validate the proposed architecture, which is built totally with open-source software, including Nginx, GeoWebCache, and MongoDB. Experimental results show that our high-concurrency WMTS can function well even when enduring more than 30,000 concurrent connections. The request throughput of the proposed high-concurrency WMTS is about twice as large as that of traditional WMTS deployed in a single web server.

Acknowledgements This work is supported by the National High Technology Research and Development Program of China (Grant: 2012AA121401) and the China Postdoctoral Science Foundation (Grant: 2012M511672).

References

Banga, G., Druschel, P.: Measuring the capacity of a web server under realistic loads. World Wide Web 2(1), 69–83 (1999)

Chiang, M. L., et al.: Design and implementation of an efficient web cluster with content-based request distribution and file caching. Journal of Systems and Software 81(11), 2044–2058 (2008)

Faour, A., Mansour, N.: Weblins: A scalable WWW cluster-based server. Advances in Engineering Software 37(1), 11–19 (2006)

Ferrari, D.: On the foundations of artificial workload design. ACM SIGMETRICS Performance Evaluation Review 12(3), 8–14 (1984)

Fisher, D.: Hotmap: Looking at geographic attention. IEEE Transactions on Visualization and Computer Graphics 13(6), 1184–1191 (2007)

Gautam, N.: Performance analysis and optimization of web proxy servers and mirror sites. European Journal of Operational Research 142(2), 396–418 (2002)

GeoWebCache. http://GeoWebCache.org/

Krishnamurthy, D., et al.: A Synthetic Workload Generation Technique for Stress Testing Session-Based Systems. In: Proceedings of the IEEE Transactions on Software Engineering 32(11), 868– 882 (2006)

Liu, Z., et al.: Traffic model and performance evaluation of web servers. Performance Evaluation 46(2-3), 77–100 (2001)

MongoDB. http://www.mongodb.org/

Open Geospatial Consortium Inc. OpenGIS Web Map Tile Service Implementation Standard, OGC 07–057r7, 2010.04

Mosberger, D., Jin, T.: httperf: A tool for measuring web server performance. ACM SIGMETRICS Performance Evaluation Review 26(3), 31–37 (1998)

Reese, W.: Nginx: the high-performance web server and reverse proxy. Linux Journal, (2008)

Sharifian, S., et al.: A content-based load balancing algorithm with admission control for cluster web servers. Future Generation Computer Systems 24(8), 775–787 (2008)

Yang, J., et al.: Modeling and simulation of performance analysis for a cluster-based Web server. Simulation Modelling Practice and Theory 14(2), 188–200 (2006)

Chapter 15
Improved Parallel Optimal Choropleth Map Classification

Jason Laura and Sergio J. Rey

Abstract In this chapter we introduce an improved parallel optimal choropleth map classification algorithm to support spatial analysis. This work contributes to the development of a Distributed Geospatial CyberInfrastructure and offers an implementation of the Fisher-Jenks optimal classification method suitable for multi-core desktop environments. We provide a description of both a single-core vectorized implementation and a parallelized implementation. Our results show that single core vectorization alone provides computational speedups compared to previous parallel implementations and that a combined, parallel and vectorized, implementation offers significant speed improvements.

Keywords Parallelization • Vectorization • Spatial analysis • PySAL

15.1 Introduction

The current trend to deploy institutional High Performance Computing (HPC) clusters composed of hundreds or thousand of computing cores and the increase in multi-core micro-compute (desktop) units provides the means by which vast quantities of spatial data can be rapidly analyzed (Wang and Armstrong 2009; Yang et al. 2008; Yang and Raskin 2009). The continual increase in spatial data size is a product of increased data capture initiatives, improved remote sensing capabilities, and increasingly complex processing models (Wang and Armstrong 2003). Spatial algorithm performance, as measured by total compute time, has not kept pace, largely due to limitations in serial, single core, processing. Current research focuses on the

J. Laura (✉) • S.J. Rey
School of Geographical Sciences and Urban Planning,
GeoDa Center for Geospatial Analysis and Computation,
Arizona State University
e-mail: jlaura@asu.edu; srey@asu.edu

X. Shi et al. (eds.), *Modern Accelerator Technologies for Geographic Information Science*,   197
DOI 10.1007/978-1-4614-8745-6_15, © Springer Science+Business Media New York 2013

development and implementation of a unified Geospatial Cyberinfastructure(GCI) that provides data storage and retrieval functionality, distributed spatial analytical tools, and a middleware layer to facilitate communication (Yang and Raskin 2009; Yang et al. 2010). A key middleware research theme seeks to address the disparity between serial spatial analytical algorithm performance and rapid increases in data volume (Wang and Armstrong 2003, 2009; Yang et al. 2008).

Research into distributed or parallel spatial algorithm implementation falls largely into two categories: theoretical or applied. Armstrong et al. (1993), Wang and Armstrong (2003), Yang et al. (2008, 2011) explore the potential benefits of distributed analytical methods as a means to process large data sets or utilize complex process models. Through distributed data storage, computation, and end-user access capabilities, it should be possible to both facilitate multi-domain collaborative research efforts and reduce or eliminate the need for data aggregation. Applied research, focusing on the decomposition and implementation of parallel spatial algorithms can be further divided into two broad temporal categories. Before the widespread adoption of multi-core desktop computers (Armstrong and Marciano 1995, 1996; Armstrong et al. 1993; Griffith 1990) develop hardware specific parallel implementations designed to run on super-computers. These implementations are largely constrained to a single piece of hardware due to implementation specifics. As multi-core desktop computers permeated the consumer market, Wang and Armstrong (2003, 2009) explore domain decomposition through the implementation of parallel spatial analytical tools and Rey et al. (2013) tests multiple programming libraries to facilitate parallelization of a single spatial algorithm. Each of these applied implementations seeks to leverage additional computational potential through the parallelization of serial spatial algorithms.

The rapid increase in data size and increasingly complex process models vastly outpace the current computational abilities of serial spatial analytical algorithms (Wang and Armstrong 2003). Anecdotally, Anselin et al. (2004) laments the inability to perform Exploratory Spatial Data Analysis (ESDA) due to programming language performance limitations and overall computational expense as the number of observations becomes non-trivial. These limitations are extensible beyond algorithms classified under ESDA and limitations correlating only to sample size. Parallelization across the entire spatial analysis stack requires knowledge of techniques to rapidly employ heuristic methods for complex combinatorial problems or iterative analytical algorithms requiring high numbers of Monte Carlo permutations (Duque et al. 2011; Rey et al. 2013; Wang and Armstrong 2003). Acceleration through parallelization and vectorization provides a means by which the spatial analysis stack can leverage the available computational power of modern desktops. This in turn provides the opportunity for researchers to avoid process generalization or areal unit aggregation to combat long runtimes.

The remainder of this chapter is organized as follows: In Sect. 15.2 we provide an introduction to parallelization, vectorization, and the use of a single shared memory space. These three implementation paradigms provide a means by which improved performance can be realized. Section 15.3 describes a technique for optimal map classification for the generation of choropleth map visualizations and is broadly segmented into three subsections. Section 15.3.1 provides the formulation of the Fisher-Jenks optimal data partitioning algorithm, 15.3.2 describes previous parallel implementation initiatives of said algorithm, and 15.3.3 introduces PySAL a spatial analytics library used to implement our work. In Sect. 15.4 we provide a description of our implementation experiment, test hardware, and results. Section 15.5 highlights the key implementation challenges and Sect. 15.6 concludes with a summary and identification of areas of future research.

15.2 Parallelization, Vectorization, and Shared Memory

Three programming paradigms provide tools to utilize all available computational power and decrease total compute time. Parallelization, or the process of distributing concurrent computation to two or more processing cores, facilitates the greatest reduction in total processing times for large data sets. Vectorization, defined below, leverages the ability for modern processors to perform multiple concurrent computations in a single processor cycle. This is a single core optimization technique that provides notable speed improvements. Finally, shared memory facilitates concurrent read and write access to a single Random Access Memory (RAM) space. This alleviates the need for in memory data duplication for processing, i.e. data available to the mother process is also available to the child processes without copying.

15.2.1 Parallelization

Parallelization is the process by which previously sequential computation is distributed over multiple Central Processing Units (CPUs) or Graphics Processing Units (GPUs) and concurrently performed. In the context of this work, parallelization is controlled by a single mother process and load is delegated to child processes. This is a hierarchy of task allocation, by the mother, and job completion, by the child. Broadly, it is possible to classify the parallelization of serial algorithms as either embarrassingly parallel, where communication between children is not necessary and synchronization is managed by the mother, or complexly parallel where the mother manages asynchronous communication between the children.

The maximum achievable speedup through parallelization is governed by Amdahl's Law. This law states that the total theoretical speedup attainable is a function of the processing that can occur in parallel and the processing that must occur in serial. Serial computational expense includes communication between cores, serial input / output (I/O), and job distribution. This law can be formulated as:

$$Speedup = \frac{1}{(1-f) + \frac{f}{n}} \tag{15.1}$$

where $1-f$ is the portion of the computation that must be computed in serial and $\frac{f}{n}$ is the portion of the program distributed over n cores (Hill and Marty 2008). Maximum speedup, attained through the application of Amdahl's Law, can be found through the parallelization of the most computationally expensive, sequential, portions of an algorithm, i.e the bottlenecks (Hill and Marty 2008). A reduction in f, in the numerator, directly impacts the attainable speed increase.

15.2.2 Vectorization

Single Instruction, Multiple Data (SIMD) processors are capable of performing vectorized computation (Flynn 1972). Vectorized computation takes multiple input elements and performs the same operation on each of the elements. For example, the addition of two vectors on a non-SIMD processor requires that each element is sequentially accessed, queried for the arithmetic operation to be performed, processed, and returned. A processor capable of vector computation can concurrently access all elements of the input data and apply a single arithmetic operation (single instruction) to all elements (multiple data). A key component of leveraging this type of computation is data representation. When possible we represent data as vectors and leverage a high level processing library, NumPy (Oliphant 2006), to handle machine level vectorization. Successful vectorization requires that the input data be a single vector, a row of a matrix, or a subset of a row; the dimensionality of the input data is key, not the overall storage strategy (Buzbee 1986).

15.2.3 Shared Memory

The goal of this research is to reduce spatial algorithm compute time and facilitate the analysis of large datasets with complex process models. Therefore, RAM provides a high speed data access and storage platform. Adam Jacobs (2009) suggests that randomly ordered access to RAM is 100,000 times faster than a traditional hard drive lookup and 100 times faster than a lookup on a solid state hard drive. In implementation this means that parallel algorithms should strive to store all required information in RAM. This requires that both the mother and child processes have access to the same data.

The distribution of data from the mother process to the child process differs based upon the underlying operating system. On POSIX compliant or nearly compliant system, data is seamlessly shared between the mother, managing I/O and all child processes (van Rossum and Drake 2013). Conversely, the Windows operating system does not dynamically share memory space between processes and therefore access to data available to the mother is not explicitly available to the children. For both system types it is possible to allocate a shared memory space, populate it with data, and then share that data either as a variable pointer, for POSIX systems or a direct to memory pointer, for non-POSIX systems. Access to a single read / write memory space for all processes removes data copy overhead and shared memory is essential to achieve the highest possible speeds with the lowest necessary memory overhead.

15.3   Optimal Map Classification for Choropleth Mapping

Considerable research effort has been applied to the application, suitability, and generation of choropleth maps (Brewer and Pickle 2002; Burrough and McDonnell 1998; Slocum et al. 2008). Choropleth maps utilize color or patterning to differentiate between areal units based upon some underlying attribute. The partitioning of attribute data for visualization is the focus of this work. Classification, or data partitioning, can take many forms including equal interval, frequency, or standard deviations from a mean (Brewer and Pickle 2002). One popular method for data segmentation employs the Fisher-Jenks optimal classification algorithm to break data into statistically derived classes such that the variation between classes is maximized and the variation within classes is minimized. This is a non-spatial data partitioning algorithm applied to spatial data.

15.3.1   Fisher-Jenks Algorithm

The Fisher-Jenks algorithm optimally classifies n observations into k classes such that all observations are members of a single class. Structured as an optimization problem, the algorithm is constrained to minimize some measure of variance within each class and maximize variance between classes. This can be the absolute sum of squares deviation from the class median or the sum of the squared deviations around the class mean (Rey et al. 2013). The Fisher-Jenks algorithm has an $O(n^k)$ runtime for unordered data and an $O(k^n)$ runtime for ordered data sets (Hartigan 1975; Rey et al. 2013). Given the runtime, computation of medium to large dataset is infeasible in a serial ESDA environment.

The Fisher-Jenks algorithm consists of three steps: (1) the computation of a diameter matrix which stores the sum of squares variance from the mean for all clusters, (2) the computation of an error matrix which stores the minimum variance of a set of n observations for k classes, and (3) the query of the error matrix to find those pivots which fulfill the optimization constraints (Hartigan 1975).

1. Compute the diameter $D_{i,j}$ for all pairs of n such that $1 \leq i \leq j \leq n$. Diameter in this work is defined as the sum of squared deviations about the mean.
2. Populate each element, L, of the error matrix for rows $[2,k]$ by $E[P_{i,L}] = min(D_{1,j-1} + E[P_{j-1,L-1}])$. This is dynamically generated as the error of the optimal partition for the current row index, $2 \leq j \leq k$, is derived from the preceding row index, $j-1$.
3. Locate the optimal partition from the error matrix as $E[P_{n,k}] = E[P_{j-1,k-1}] + D_{j,n}$

We reduce the total number of steps to three by populating the first row of the error matrix from the first row of the diameter matrix. This is in contrast to the original publication by Hartigan (1975) and subsequent work by Rey et al. (2013) which describe a fourth step, occurring between steps one and two to populate the first row of the error matrix. Additionally, previous works implemented this algorithm either in serial, or through the parallelization of step one. We parallelize both steps one and two.

15.3.2 Previous Work

Rey et al. (2013) implemented a parallel Fisher-Jenks algorithm using three freely available parallel Python libraries: the built-in Python module multiprocessing, Parallel Python, and PyOpenCL. Language syntax requirements differ between each library and therefore require divergent implementations. The built-in multiprocessing module ships with Python versions greater than 2.6, offers shorter development times due to more straightforward implementation requirements, and reported the best results. Parallel Python, an external library, requires an additional user installation and reported the worst parallel performance. Finally, PyOpenCL, a library designed to leverage either the Central Processing Unit (CPU) or Graphics Processing Unit (GPU), requires an additional installation step, complex implementation requirements, and returned repeated memory allocation errors. In light of these results, our implementation utilizes multiprocessing.

In addition to implementing and testing three libraries (Rey et al. 2013) offer multiple insights into the parallelization and implementation of the Fisher-Jenks algorithm. These insights have been utilized to drive our research to further improve algorithm portability and performance. First, in-memory duplication limited total sample size to half of the available RAM space. Second, the parallel computation of the diameter matrix improved total compute time sufficiently that the computation of the error matrix is revealed as a new bottleneck. Third, parallelization is only beneficial with medium to large sample sizes as costs associated with parallel overhead must be accounted for.

15.3.3 PySAL

PySAL, an open-source spatial analytics library, written entirely in Python provides the test platform for our work (Rey and Anselin 2010). Being open-source, the code is freely available to end users for modification, exploration, and usage. This open source model provides transparency in implementation as well as the ability to integrate externally coded contributions from non-core developers. Additionally, rapid and iterative code generation is a hallmark of Python and a driving reason behind the creation of PySAL using this language (Rey et al. 2013). This design philosophy extends to our implementation, which is undergoing integration into the core of

PySAL. We utilize the core implementation of the serial Fisher-Jenks algorithm and a parallel branch which includes an implementation using the Python built-in library multiprocessing to benchmark our code.

15.4 Implementation Specifications

Our improved implementation focuses on three extensions of the work by Rey et al. (2013). First, we explore the ability to avoid in-memory data duplication through the use of shared memory space. Second, we refractor the computation of the diameter matrix to leverage vectorized computation. Finally, we parallelize and vectorize the computation of the error matrix. After implementing these changes we test a range of sample sizes (n) and classes (k) to compare our results to both the serial Fisher-Jenks implementation and the initial parallel implementation.

15.4.1 High Level Parallel Implementation

Recall from above the Fisher-Jenks algorithm consists of three phases: computation of the diameter matrix, computation of the error matrix, and identification of the data pivot points. We parallelize phases one and two of the algorithm. We conclude that the parallelization of the pivot point computation, phase three, is not necessary as total compute times remain less than 0.5 % of total compute time and decreases with increases to n.

15.4.2 CTypes Shared Memory

Prior to initiating the three algorithm phases, described above, we initialize all the necessary data structures. This alleviates the need for in memory data duplication (Rey et al. 2013). Accomplishing this requires that two contiguous memory blocks be pre-allocated using the built-in ctypes library (van Rossum and Drake 2013). This library provides a Pythonic interface to non-local functions and, in our usage, facilitates the access of a single globally available memory space by all child processes. In this context, in RAM storage must either be allocated at the largest possible data type, 64-bit floating point, or the input data must be sampled and the data type intelligently determined. We utilize the former approach in our implementation. Finally, this implementation can be classified as embarrassingly parallel and concurrent writes are not required. Therefore, the ctypes allocated memory does not have accompanying locking mechanisms (locks or semaphores).

Two constraints and two benefits are introduced through the use of shared memory. First, access to shared memory using Python requires the use of pointers to a memory address; this is not direct in-language access to the stored elements. It is therefore necessary to read directly from the memory buffer. This is accomplished through the use of the *from buffer* () function within NumPy (Oliphant 2006). The second constraint requires that the buffer is stored as a flat array, (i.e. one dimensional). The Fisher-Jenks algorithm requires that the diameter matrix be *n x n* and the error matrix be *n x k*. Therefore, it is necessary to reshape the buffer view before in-language processing. While it is not possible to make a pythonic view of the shared memory space globally available to all children on a non-POSIX system, it is possible to pass the pointer to a shared memory space and then recapture a pythonic view without issue. In this manner, the use of ctypes facilitate OS portability in a manner that the use of global shared memory, internal to python, would not.

The pre-allocation phase concludes with packing each row of the diameter matrix with the sorted input values. This is to facilitate the vectorization of the computation in the next phase.

15.4.3 Diameter Matrix Computation

Generation of the diameter matrix is the most computationally expensive portion of the Fisher-Jenks algorithm and Rey et al. (2013) show that the parallelization of this phase provides non-trivial speed increases for medium and large problem sets. Our implementation follows theirs, but differs in two ways. First, we avoid the use of memory duplication by passing a pointer and row indices from the mother to the child process. Second, we remove all *for* loops from the code to leverage vectorization in the computation of the diameter matrix. These two general improvements provide substantial speed improvements over the initial parallel implementation and are more explicitly described below.

Diameter matrix computation is initiated with the mother process computing the load for each core using the equation

$$interval = \frac{n}{c}, \tag{15.2}$$

where n is the total number of values and therefore rows, and c is the number of cores. Data decomposition using this method often leaves excess rows that are processed by the first core to complete its initial load. Once the segmentation of the load is computed a memory pointer and the indices of the rows to be processed are passed to each child process. Once distributed, each core is assigned $\dfrac{n}{numbercores}$ rows.

Each child process then iterates over its assigned rows and computes each row using vectorized computation that fully leverages the SIMD capabilities of the processor. Due to the fact that the lower triangle of the matrix is zero, before processing a row we first replace the i elements with zero where i is the row number. Once pre-processing of the row is completed, we compute the entire diameter matrix row with a single operation, i.e. same instruction multiple data, using the scalar equation

$$D_{I,J} = \sum_{i=I}^{J} (y_i - \bar{y}_{I,J})^2 \tag{15.3}$$

where $D_{I,J}$ is the diameter of the cluster consisting of values I through J, $\bar{y}_{I,J} = \dfrac{1}{J - I + 1} \sum_{i=I}^{J} y_i$, and y_i is the attribute value for observation i. Each row of the matrix D is obtained through vectorization.

Once each child has completed the assigned load the jobs synchronize and the mother process initiates computation of the error matrix. In all phases the mother process manages synchronization, but also acts as a child process, performing a segment of the total load. This functionality exists within the Python language without programmer implementation.

15.4.4 Error Matrix Computation

Unlike Rey et al. (2013) we also parallelize the computation of the error matrix. This is a direct extension to earlier work as computation that was sufficiently fast previously now becomes the primary processing bottleneck. The first step of this phase is to copy the top row of the diameter matrix to the top row of the error matrix. This reduces the total number of computations from k^n to $(k-1)^n + 1$.

The computation of the error matrix is decomposed differently than the diameter matrix. Instead of sending complete rows to each child process, the decomposition technique used for the diameter matrix, it is necessary to send segments of a single row to each process. This is because each row of the error matrix depends upon values from both the diameter matrix and the preceding row of the error matrix. Therefore, we distribute the computation of each row over each available core using the process summarized in Table 15.1. Here e is an element in the error row, c is a child process (core), n is the total number of values, and m is a total count of the available cores. As the row index increases the total computational load increases, but the computational load required to intelligently distribute the load exceeds the total computational cost of this phase.

Once row segmentation is computed, the mother process distributes memory pointers and row indices to each child process, as above. We then compute each error element as the minimum of the sum of elements from the preceding error matrix row and elements from a column of the diameter matrix. Both sequences can be represented as vectors and therefore provide a means to performed vectorized computation using the following equation

$$\begin{aligned}
E_{i,j} &= [e_{i-1,j-1}, \quad e_{i-1,j-2}, \quad e_{i-1,j-3}, \quad \cdots \quad e_{i-1,(j-n)+2}, \quad e_{i-1,(j-n)+1}] \\
D_{i,j} &= [d_{i,j}, \quad d_{i+1,j}, \quad d_{i+2,j}, \quad \cdots, \quad d_{(i+k)-1,j}, \quad d_{i+k,j}] \\
e_{i,j} &= \min(E_{i,j} + D_{i,j}),
\end{aligned} \tag{15.4}$$

Table 15.1 Segmentation of the error matrix over available cores

Core number	Error row segment (vector)					
c_1	$e_{i,1}$	$e_{i,2}$	$e_{i,3}$	$e_{i,4}$	$\cdots$	$e_{i,\frac{n}{c}}$
c_2	$e_{i,\frac{n}{c}+1}$	$e_{i,\frac{n}{c}+2}$	$e_{i,\frac{n}{c}+3}$	$e_{i,\frac{n}{c}+4}$	$\cdots$	$e_{i,2*(\frac{n}{c})}$
$\vdots$	$\cdots$	$\cdots$	$\cdots$	$\cdots$	$\cdots$	$\cdots$
c_m	$e_{i,m*(\frac{n}{c}+1)}$	$e_{i,m*(\frac{n}{c}+2)}$	$e_{i,m*(\frac{n}{c}+3)}$	$e_{i,m*(\frac{n}{c}+4)}$	$\cdots$	$e_{i,n}$

where $E_{i,j}$ is a vector extracted from the previous row of the error matrix, $D_{i,j}$ is a vector extracted from the diameter matrix, and $e_{i,j}$ is the minimum scalar element of $E_{i,j}+D_{i,j}$. While we still must iterate over each index in the error matrix, it is possible to leverage the SIMD capabilities of the processor to populate each error index.

15.4.5 Pivot Matrix Computation

Finally, we find the pivot indices or values in the error matrix such that the variance is maximized between classes and minimized within classes. This is an extremely fast lookup that is performed in serial. The implementation is identical to that of Rey et al. (2013), except that our underlying data structure is an array instead of a list. This incurs a negligible performance hit to our implementation less than 0.5 % of total compute time.

15.4.6 Experiment, Hardware, and Results

Below we report the results of testing the improved parallel implementation against both the implementation created by Rey et al. (2013) and the original serial implementation in PySAL, which mirrors (Hartigan 1975). To control for hardware variation we report comparative results from a single machine after performing a clean reboot. To test the impact of parallelization of both the diameter matrix and error matrix computation we test $k = 5,7,9$, which are reported to be the most commonly selected numbers of classes (Rey et al. 2013) and a range of sample sizes from $n=1000$ to $n=12,000$, incrementing in intervals of 2000 once that threshold is attained. For tests including the original serial implementation the maximum value tested is $n=8,000$ due to excessive runtimes. The test data is randomly generated floating point numbers with a range of $(0,1]$. Finally, we test $n>12,000$ on a server level machine to extend the performance curve and explore the current upper bounds of this implementation.

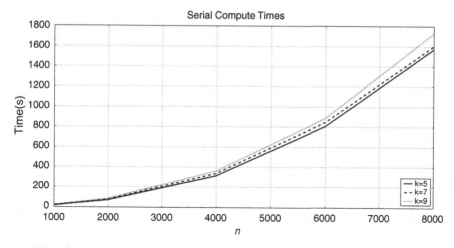

Fig. 15.1 Piecewise linear speed curve showing total compute time of the serial Fisher-Janks algorithm

15.4.7 Hardware

The test hardware consisted of an Intel 3.1GHz i3-2100 Sandy Bridge dual core processor, that reports as 4 cores due to hyper-threading (Intel 2003), with 4GB of RAM, running KUbuntu linux. This is a pseudo POSIX compliant system that offers processor level process forking and mirrors a low end machine users are likely to find readily accessible. For $n > 12{,}000$ we utilize a 12-core 2.26 GHz Mac Pro with 64 GB of RAM.

15.4.8 Results

A single-core, serial Fisher-Jenks implementation provides a benchmark against which it is possible to compute the total speedup attained through parallelization. Figure 15.1 shows the piece-wise linear compute time curves generated by the serial algorithm which clearly grow with the sample size. Additionally, the number of classes increases total compute time, but this impact is small when compared to the correlation between n and t, the total compute time.

To compare the results speedup results we utilize the standard speedup curve formulated as

$$speedup = \frac{S_{n,k}}{P_{n,k}}, \qquad (15.5)$$

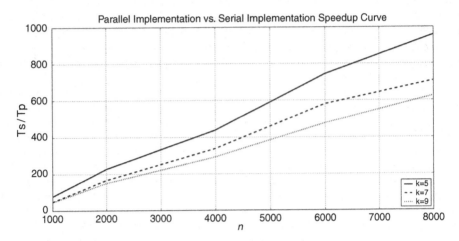

Fig. 15.2 Speedup curve benchmarking the serial implementation to our parallel implementation

where $S_{n,k}$ is the serial computation time as a function of the number of samples, n, and the number of classes k, and $P_{n,k}$ is the total parallel computation time given identical n and k values.

Figure 15.2 illustrates the speedup attained compared to the original serial implementation. We find that for $n > 1,000$ the overhead associated with parallelization is significantly less than the total speedup attained. This is in-line with previous results (Rey et al. 2013). The speedup curves are largely linear and clearly vary with k. This is expected as increases in k introduce both an additional k^n computations and the associated parallelization overhead. Unlike, Rey et al. (2013) we do not find a plateau at $n = 2,000$. Finally, we find total speedups ranging from 50 times faster to nearly 1,000 times faster.

Moving to a comparison of our implementation to the previous parallel implementation (Rey et al. 2013), we again show, in Fig. 15.3, a general speed increase between 25 times and 200 times faster. This is largely attributable to the reduction in in-memory duplication, the use of vectorization, and the parallelization of computation for the error matrix. Interestingly, we see a plateau and overall decrease in speedup from $n = 8,000$ to $n = 12,000$. This is potentially a product of naive data decomposition for error matrix computation and additional tests comparing larger numbers of samples are required. Finally, we also see a marked improvement comparing our single core, vectorized, algorithm to the previously published multi-core algorithm (Rey et al. 2013). Vectorization alone provides speedups of between twenty-five and fifty over non-vectorized multi-core implementations.

Figure 15.4 compares the speed gains attained by a solely vectorized implementation and the final implementation leveraging both vectorization and parallelization. Clearly the later provides greater speed increases, but the total difference between implementations is negligible until $n > 1,000$. Given the hardware specific requirements inherent to leveraging all available processing cores, and the human time required to implement a parallel implementation, we suggest that single core vectorization may provide implementations which are sufficiently fast. This must be assessed

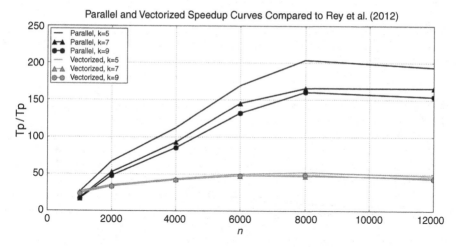

Fig. 15.3 Speedup curve comparing the original and improved parallel implementations

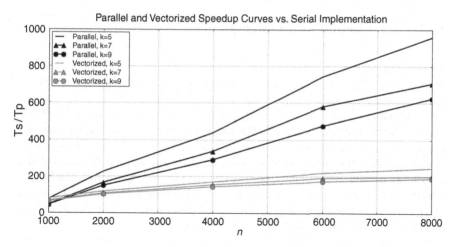

Fig. 15.4 Speedup curve benchmarking comparing both vectorized and parallel implementations to the serial implementation

on a problem specific basis. For $n < 1,250$ we see that vectorization out performs parallelization; this is due to the overhead associated with spawning child processes.

Finally, Fig. 15.5 depicts benchmarking performed on the server level machine to compare total computational time for the parallel and vectorized implementations. Tests were performed from $n = 1,000$ to $n = 42,000$ and show total compute time leveraging both parallelization and vectorization for large problem sets remains well under two minutes. We can report a piece-wise linear function for solely vectorized computation. Additional testing with larger values of n is required to classify the expected behavior of the parallel implementation speed curve.

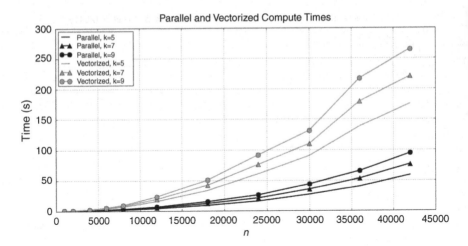

Fig. 15.5 Total computation of the vectorized and parallel implementations

15.5 Implementation Challenges

The development of an improved parallel Fisher-Jenks algorithm was an iterative process encountering multiple implementation challenges and identifying opportunities for future work. First, refactoring the original Fisher-Jenks algorithm to allow for vector representation in the computation of the diameter and error matrices was human time intensive. This required that the problem be recast and represented in a completely different structure. Second, attempting to scale this work for big data is an ongoing challenge as the problem grows quadratically in not only time, but also in memory consumption. Finally, porting this code from a POSIX to a non-POSIX system required an additional refactoring of the shared memory space and exploration of efficient means to pass access to shared memory between processes which do not exist in the same variable space (namespace).

15.6 Extensions and Future Work

This extension to Rey et al. (2013) highlights future research objectives and provides additional insight into deploying parallel algorithms throughout the spatial analysis stack. First, using open source, built-in libraries, it is possible to develop and deploy system agnostic, parallel, code. This requires that code be developed understanding the limitations placed by each of the three modern desktop operating systems. Second, we concur with Rey et al. (2013) in that speed improvements attained through parallelization are valid only for medium to large values of n. This is in-line with expectations as process forking and inter-core communication incur an overhead that is non-trivial above a threshold. Vectorization provides a method by which increased performance can be attained for small values of n.

Through this work, we can highlight the following three insights into the parallelization of this algorithm. First, the representation of data as regular arrays, when possible, is essential to providing the means by which vectorization can occur. Major performance gains are attainable without parallelization, i.e. in serial, when the underlying vector computation capabilities of the processor can be leveraged. Additional research is required to determine the best methods to represent jagged arrays, i.e. non standard number of columns per row, and still leverage vector computation. Second, when refactoring for a high level parallelization implementation, it is necessary to iteratively deploy code and highlight processing bottlenecks at each iteration. This is evidenced by the performance gains attained by leveraging and improving computation of the distance matrix (Rey et al. 2013) as well as parallelizing the computation of the error matrix. This highlights the necessity for an iterative approach to code parallelization as individual bottlenecks may not exist prior to refactoring. Third, this algorithm is still memory constrained and the parallelization of spatial algorithms to improve performance must focus on both overall computational speed and efficient data representation.

Future work will focus on a low-level parallelization whereby the algorithm undergoes structural changes. This implementation is still limited by n as the total memory footprint increases quadratically with n. As HPC clusters often limit the total amount of available RAM per node and desktop computers with large quantities of RAM are not normally deployed, a low level parallelization that concurrently computes, utilizes, and overwrites the diameter matrix is essential. This implementation could theoretically scale without limit to RAM as a row in the diameter matrix would exist only long enough to derive the associated error matrix elements. This would be an asynchronous implementation requiring inter-core communication and dynamic load balancing.

Additional future work is also required to explore porting this implementation to a HPC cluster. The extensive use of shared memory space requires that all cores, and by extension nodes, access the same data. In some HPC environments this level of shared-memory is not available and inter-node communication, which is slower than RAM access, is required. The potential exists that a low level parallelization, as outlined above, could alleviate this issue, or another refactoring of the algorithm is required.

Finally, this work is a single parallelization of a single algorithm within the spatial analysis stack. To fulfill the goal of providing a taxonomy of parallel spatial algorithm implementation methodologies additional parallelization techniques, of divergent spatial algorithms classes must be undertaken. This work should be broad in scope and provide documentation of successful and unsuccessful parallelization efforts specific to the domain of spatial analysis.

Acknowledgements This research was funded in part by NSF Award OCI-1047916, SI2-SSI: CyberGIS Software Integration for Sustained Geospatial Innovation. We thank the anonymous referees and the editors for their constructive comments.

References

L. Anselin, Y. W. Kim, and I. Syabri. Web-based analytical tools for the exploration of spatial data. *Journal of Geographical Systems*, 6(2):197–218, June 2004.

M.P Armstrong and R. Marciano. Massively parallel processing of spatial statistics. *International Journal of Geographical Information Systems*, 9(2):169–189, 1995.

M.P. Armstrong and R. Marciano. Local Interpolation Using a Distirbuted Parallel Supercomputer. *International Journal of Geographical Information Systems*, 10(6):713–729, 1996.

M.P Armstrong, C.E. Pavlik, and R. Marciano. Parallel Processing of Spatial Statistics. *Computers & Geosciences*, 20(2):91–104, 1993.

C. A Brewer and L. Pickle. Evaluation of methods for classifying epidemiological data on choropleth maps in series. *Annals of the Association of American Geographers*, 92(4):662–681, 2002.

P.A. Burrough and R. McDonnell. *Principles of geographical information systems*. Oxford University Press, 1998.

B.L. Buzbee. A Strategy for Vectorization. *Parallel Computing*, 3:187–192, 1986.

J.C. Duque, R.L. Church, and R.S. Middleton. The p-Regions Problem. *Geographical Analysis*, 43(1):104–126, January 2011.

M.J. Flynn. Some Computer Organizations and Their Effectiveness. *IEEE Transactions on Computers*, C-21(9):948–960, September 1972.

D.A. Griffith. Supercomputing and Spatial Statistics: A Reconnaissance. *The Professional Geographer*, 42(4):481–492, 1990.

J. A. Hartigan. Partition by Exact Optimization. In *Clustering Algorithms*, chapter 6, pages 130–142. Wiley, New York, New York, USA, 1 edition, 1975.

M.D. Hill and M.R. Marty. Amdahl s Law in the Multicore Era. *Computer*, 41(7):33–38, 2008.

Intel. Intel Hyper-Threading Technology. Technical Report January, Intel Corporation, 2003.

Adam Jacobs. The pathologies of big data. *Communications of the ACM*, 256:1–12, 2009.

T.E. Oliphant. *Guide to NumPy*. Provo, UT, March 2006.

S.J. Rey and L. Anselin. PySAL: A Python library of spatial analytical methods. In A. Fischer, M.M ; Getis, editor, *Handbook of Applied Spatial Analysis*, pages 175–193. Springer, 2010.

S.J. Rey, L. Anselin, R. Pahle, X. Kang, and P. Stephens. Parallel optimal choropleth map classification in pysal. *International Journal of Geographical Information Science*, pages 1–17, 2013.

T. Slocum, R. McMaster, F. Kessler, and H. Howard. *Thematic cartography and geovisualization*. Prentice Hall., 2008.

G. van Rossum and F.L. Drake. Python Reference Manual, 2013.

S. Wang and M.P. Armstrong. A quadtree approach to domain decomposition for spatial interpolation in Grid computing environments. *Parallel Computing*, 29(10):1481–1504, October 2003.

S. Wang and M.P. Armstrong. A theoretical approach to the use of cyberinfrastructure in geographical analysis. *International Journal of Geographical Information Science*, 23(2):169–193, February 2009.

C. Yang, W. Li, J. Xie, and B. Zhou. Distributed geospatial information processing: sharing distributed geospatial resources to support Digital Earth. *International Journal of Digital Earth*, 1(3):259–278, September 2008.

C. Yang and R. Raskin. Introduction to distributed geographic information processing research. *International Journal of Geographical Information Science*, 23(5):553–560, May 2009.

C. Yang, R. Raskin, M. Goodchild, and M. Gahegan. Geospatial Cyberinfrastructure: Past, present and future. *Computers, Environment and Urban Systems*, 34(4):264–277, July 2010.

C. Yang, H. Wu, Q. Huang, Z. Li, and J. Li. Using spatial principles to optimize distributed computing for enabling the physical science discoveries. *Proceedings of the National Academy of Sciences of the United States of America*, 108(14):5498–503, April 2011.

Part VI
Vision and Applicability of MAT for Geospatial Modeling and Spatiotemporal Data Analytics

Chapter 16
Pursuing Spatiotemporally Integrated Social Science Using Cyberinfrastructure

Xinyue Ye and Xuan Shi

Abstract The rich details of space-time complexity in social science remain largely unexplored because of the challenge of intensities of data and computing. The current space-time simulation and statistics for social science research can only deal with a limited amount of data. We introduce a pilot study about how to deploy the modern accelerator technology and hybrid computer systems to extend the National Institute of Justice-funded Near-repeat calculation, a typical social science application? This pilot study demonstrates that it is promising to leverage high performance computing for solving large-scale space-time interaction problems, which has long been a challenging statistical issue for spatiotemporally integrated social science.

Keywords Space-time complexity • Social science • Near-repeat • High performance computing • Spatiotemporally integrated social science

16.1 Introduction

According to Waldo Tobler, "everything is related to everything else, but near things are more related than distant things" (Tobler 1970). Ignoring this relationship may lead to overlooking many possible interactions and dependence among space, time, and attributes in decision-making and policy-generation. As a result, spatiotemporally differentiation could be inevitably uneven when the same policy can be implemented or perceived differently over space-time and by different persons, socioeconomic or partisan groups.

X. Ye (✉)
Department of Geography, Kent State University, Kent, OH 44242, USA
e-mail: xye5@kent.edu

X. Shi
Department of Geosciences, University of Arkansas, Fayetteville, AR 72701, USA
e-mail: xuanshi@uark.edu

X. Shi et al. (eds.), *Modern Accelerator Technologies for Geographic Information Science*, 215
DOI 10.1007/978-1-4614-8745-6_16, © Springer Science+Business Media New York 2013

Spatial thinking has become a common interest in a growing research community of spatiotemporally integrated social science, aiming at analyzing spatial patterns of socioeconomic trends and the dynamics of geographical structures, in order to harness large volumes of digital socioeconomic data towards improving human well-being (Goodchild 2009; Ye and Rey 2011). There is growing consensus that many socioeconomic dynamics are spatiotemporally concentrated Since resources are always limited, socioeconomic development would have been concentrated in different regions but evolving in accordance with the changing market and environment. While place-based public policy has been effective (Ye and Carroll 2011a), location-based information is difficult and expensive to retrieve (Batty 2010). At the same time, researchers have recognized that spatial and temporal components of events should be addressed jointly instead of being treated separately (Andrienko and Andrienko 2006; Goodchild 2009; Anselin 2012; Ye and Liu 2012). Researchers thus have to pay more attention to both spatial and temporal aspects in thematic research (Andrienko and Andrienko 2012). For example, economic events such as massive layoffs are highly correlated across space and over time (Ye and Carroll 2011b). While many socioeconomic theories cover the space-time dimensions and there is an increasing awareness of its importance in the empirical analysis, the rich details of space-time complexity remain largely unexplored because of the constraint in computation capacity in response to the challenge of data and intensive computing. As a result, the current space-time simulation and statistics for social science research can only deal with a limited amount of data.

Timely and rigorous analysis of emerging socioeconomic events will open up a rich empirical context for the social sciences and policy interventions (Ye and Carroll 2011a). It is critical and helpful to understand the impact of and response to certain socioeconomic policies and events in a timely fashion (Warf and Sui 2010). Although integrated space-time analysis has the potential to provide an unprecedented opportunity to researchers in studying the socioeconomic dynamics, it raises the research challenges on data intensive computation (White House 2012). The current space-time simulation and statistics for social science research, however, can only deal with a limited amount of data. As a result, the rich details of space-time complexity remain largely unexplored partially due to the constraint in computation capacity in handling large scale data (Guo and Mennis 2009; Andrienko and Andrienko 2012).

Recent advancements in environmental criminology revealed repeat and near-repeat phenomena among shootings. More specifically, the location at which a shooting occurred and its immediate surroundings face an increased risk of experiencing subsequent shootings within a fairly short period of time (Ratcliffe and Rengert 2008; Wells, Wu, and Ye 2012). This finding has direct policy relevance because it provides cues for preventing possible subsequent shootings and can be used to direct interventions such as increased patrol activities.

Exploring the potential of analyzing multi-level interaction among event data to understand the socioeconomic dynamics could lead to transformative concepts and methodologies to advance the socioeconomic research. We introduce a pilot study about how to deploy the modern accelerator technology and hybrid computer architecture and systems to extend the National Institute of Justice-funded Near-repeat calculation formulated by Ratcliffe (Ratcliffe and Rengert 2008). The near-repeat calculator is used to determine whether significant repeat and near-repeat patterns exist before and during the intervention at the city and district levels. This software combines a revised Knox test (Knox 1963, 1964) and Monte Carlo simulation process to detect near repeat crime phenomenon. The revised Knox test divides space (0–d) and time (0–t) into a number of bands indicating various spatial and temporal constraints, such as from 0 to d1, from d1 to d2, from d2 to d3, and more than d3; from 0 to t1, from t1 to t2, from t2 to t3, and more than t3. Except for the last band for space and time, all of the bands have the same bandwidth. All pairs incidents can be placed into a category that combines those spatial and temporal bands (e.g., between d2 and d3 and between t1 and t2). In other words, the number of incident pairs in each space-time band can be identified.

A Monte Carlo simulation approach is used to randomly impute the times of incidents, while holding their locations constant, because incidents are not assumed to be independent. This assumption is based on previous research that reports significant spatial clustering of crime. Each simulation generates a new value of the number of incident-pairs (simulated number) in each space-time band. With many simulations for each space-time band, all derived values form a distribution that reflects the expected distribution under a null hypothesis of no spatial-temporal relationship. This simulation makes it possible to calculate the number of observed events (i.e., shooting pairs) in each bandwidth differs from the simulated number of events. For example, if 999 simulations are run and the observed number in a specific space-time band exceeds the simulated numbers 989 times, the significant level is $1-[989/(999+1)]=0.011$. That is, the significant level is very low (0.011) so the null hypothesis is less likely to be true. Put differently, there is significant space-time clustering in this specific space-time band at the 0.05 level.

Permutations are part of a numerical approach to calculate the statistical significance of the observed value. In other words, it is used to determine how likely it would be to observe the values of an actual distribution under conditions of spatial randomness. To derive the relatively stable statistical significance, 999 simulations are usually adopted. This has a major impediment consisting of the computational bottlenecks encountered when carrying out simulations for large data sets. N incidents will generate $N \times (N-1)/2$ pairs for investigation at each single run splitting into multiple bands defined by space and time, with an additional 999 simulated runs to get the statistical significance for each band. When N reaches 30,000 for example, the calculation goes beyond the capacity of Ratcliffe's Near-repeat Calculator and computer memory. Traditional solution over desktop computer could hardly accomplish the calculation on such a scale of data. When a single GPU on desktop machine is used, we accomplish the calculation and permutation in about 48.5 min. When 100 GPUs are used in Keeneland, we accomplish the calculation and permutation in about 4 min. Building essential capability for scalable analytics on socioeconomic trends and the dynamics of geographical patterns and events using the Cyberinfrastructure will be critical to enable the research community affiliated in the social, behavioral, and economic sciences with the capability to process large scale of data in affordable time frame in order to advance the research in spatiotemporally integrated socioeconomic science.

16.2   Research Context

The availability of large-size geo-referenced event data sets along with the high performance computing technology has raised the fundamental challenges and opportunities to mainstream social science research on whether and how these new trends in data and technology can be utilized to collect thematic information and help understand the socioeconomic dynamics, in order to enhance and advance the research in social, economic, and behavior sciences in general. A variety of themes can be addressed in the research of spatiotemporally integrated socioeconomic studies. Socioeconomic dynamics can be better mapped, detected, and understood through synthesizing and analyzing event data (Warf and Sui 2010).

There is a growing consensus that many socioeconomic dynamics is spatially concentrated, such as unemployment and violence. Since resource is always limited, it should have been targeted in these concentrated regions. This place-based public policy has been effective in reducing the impacts of unemployment/crimes as well as the related social and family problems associated with them. Most recently, researchers have recognized that spatial and temporal components of crimes should be addressed jointly instead of being treated separately. Researchers are thus beginning to pay more attention to both spatial and temporal aspects of crimes. The discovery of space-time interaction among these events indicates that locations at which unemployment/crimes have occurred and their geographical surroundings face an elevated risk of experiencing follow-up layoffs/crimes in the immediate aftermath of the initial layoff/crimes.

Traditional cartographic methods and legacy spatial analysis tools might need further improvement for mapping large amounts of event data. Mapping and analyzing event data requires an innovative design of visualization methods and spatial analysis algorithms. These research challenges may facilitate a paradigm shift in the discipline of geography because geographers and cartographers are facing a crisis about how to represent this new form of information. Throughout the course of this research, we will build capability in both data intensive computing and intelligent data mining through improved near repeat analysis. This research will build, enhance and expand the community including researchers and participants from broader domain science, governmental and non-profit agencies, who share the common interest in analyzing events. Meanwhile, spatiotemporal dynamics has been a central theme to GI Scientists (Goodchild 2004).

Recent advancements in environmental criminology revealed repeat and near-repeat crime phenomena whereby a victimized location and its immediate surroundings face an increased risk of a subsequent crime for a short period of time. The first shooting is an initiator, the follow-up shooting at the same location and close in time with the initiator is a repeat, and the one close in space and time with the initiator is a near-repeat (Morgan 2001). Both repeat and near-repeat phenomena have direct implications for preventing possible follow-up shootings.

Repeat victimizations have been characterized by Skogan as "probably the most important criminological insight of the decade." He claims that the "pilling up of repeat multiple victimization is mostly what makes a high-crime neighborhood a high-crime neighborhood" (Brady 1996, p. 3). Trickett et al. (1992) explain a feasible policy implication of this research: "if high crime rates occur because of repeat victimization, crime prevention should correspondingly focus on preventing people who have already been victimized from being victimized again" (p. 81).

Evidence of repeat and near-repeat phenomena has primarily been derived from research on burglaries (Bennett and Durie 1999; Bernasco 2008; Bowers and Johnson 2004, 2005; Grubesic and Mack 2008; Johnson and Bowers 2004; Johnson et al. 2007; Johnson et al. 2009; Sagovsky and Johnson 2007; Townsley et al. 2000). Johnson and Bowers (2004), for instance, find that "a burglary event is a predictor of significantly elevated rates of burglary within 1-2 months and within a range of up to 300-400 meters of a burgled home" (p. 250). Researchers are beginning to examine these phenomena among gun violence. In Philadelphia, Ratcliffe and Rengert (2008) find that the risk of a near-repeat shooting within 14 days and 400 ft. of a previous shooting was elevated 33 % compared with the normal risk background. Moreover, Wells et al. (2012) find significant and meaningful patterns for both repeat and near-repeat shootings in Houston.

Repeat and near-repeat patterns of crime are evident, but theoretical explanations and empirical tests are underdeveloped (see Maguire et al. 2008). Researchers have proposed two rival hypotheses to explain these patterns among burglaries (Bowers and Johnson 2004; Sagovsky and Johnson 2007). More specifically, the "flag" thesis suggests that some premises have unique characteristics that attract many opportunistic burglars, while the "boost" thesis states that a subsequent burglary is dependent on or substantively related to the first. A typical example of the "boost" thesis is that the same offender(s), based on the experience gained during the first offense, will commit subsequent burglaries at the same dwelling or nearby dwellings (Townsley et al. 2000). Research is just beginning to offer preliminary evidence to support the boost thesis by using police detection data where the identities of burglars are known (Bernasco 2008). Both studies find that substantial percentages of repeat and near-repeat burglaries involve the same offender(s).

For example, incident A has a near-repeat follow-up with incident B, while incident B has a near-repeat follow-up with incident C (suppose incident C is also a near-repeat follow-up with incident A). The near repeat calculator can identify three space-time pairs: incident A and incident B, incident B and incident C, incident A and incident C. The shooting chain among the three incidents (A, B, and C) fails to be detected because only pairs are examined. However, it would be more informative to identify the three incidents as a "space-time chain" instead of three space-time pairs. It will be more useful to identify the concentration of multiple incidents than a concentration of two incidents. Detecting hot spots of violence can lead to more effective police patrol deployments and can be used to evaluate policing strategies, because limited policing resource can be directed the concentration of crimes. The criterion used to identify a hot spot is the spatial concentration of events. However, this criterion clearly misses the temporal dimension. This study suggests integrating the merits of near-repeats (space-time interaction) and hot spot. Nevertheless, when the scale of event goes from three to more, the computation challenge emerges. A method for assessing space-time chain has three parameters: spatial band, temporal band, and the number of the incidents (at least three incidents). This serves a first step to better our understanding of the role of space and time among a series of events.

From the methodological perspective, this method further implements the idea of temporal hot spots, by combing near repeat analysis and hot spot detection methods. In other words, a temporal hot spot is formed by multiple related events (each event has location and occurrence time), instead of only based on the locations of events. Hence, this tool can be used to examine the effects of hot spot policing from the policy perspective. Since the location and time of each shooting is available, the spatial and temporal distances between any two incidents can be measured. The distances are then divided into four groups regarding the proximity of incidents: close in space and close in time; close in space and not close in time; not close in space and close in time; not close in space and not close in time. The definitions of proximity and "series" will be based on a predefined space-time band and number of incidents. Let's imagine the following scenario: many local police officers in a city said that they were very concerned about series of shootings. According to them, these shootings have the following features: at least three shootings occurred within two weeks, and the spatial distance among any two shootings is less than one mile. To find out whether such patterns exist and whether they are statistically significant, intensive simulation is expected. The significant existence of paired shootings does not mean the significant existence of multiple shootings in the same space-time band. Otherwise, all the initiators or follow-up shootings should be close to each other and these shootings should form one spatial cluster in a city. However, it is far from the truth.

First, for each shooting, we can identify some possible follow-up shootings which are within its spatial-temporal band, and a filtering technique can be applied to retrieve a shooting list. This filtering technique is the core of the chain calculator.[1] Multiple shootings need to be retrieved to form a space-time chain (shooting list). This filtering technique is designed as follows: the most immediate (by time order) follow-up shooting will be selected as the second shooting in the list. A third shooting will also be chosen by time order, but at the same time it needs to be within one mile to both the first and second shootings. If this spatial criterion cannot be met, the next shooting by the time order will be chosen and the spatial distance between the candidate shooting and each of the shootings in the list will be calculated until both spatial and temporal criteria are met. This filtering is applied until the shooting list has three incidents or no candidate is available. Hence, the final shooting list might not include all the original follow-up shootings. However, this procedure guarantees that any two shootings in the list are within two weeks and one mile. In other words, the final shooting list will have three shootings which cluster in the defined space and time band. The number of all possible lists is counted for each shooting (the initiator of the space-time chain). This number is then summed, which is the number of observed space-time chains for

---

[1] The near-repeat calculator uses a simpler filtering technique. The near-repeat calculator filters out the shootings which are either not close in time or not close in space to the focal shooting.

all the shooting incidents. Finally, a Monte-Carlo test is used to evaluate whether the number of observed space-time chains is significantly larger than expected on the basis of a random distribution of shootings in space and time.

16.3   Implementation of Near-Repeat Calculation

16.3.1   Near-Repeat Calculation by Python

The current version utilizes NumPy, the fundamental package for scientific computing with Python. It contains a powerful N-dimensional array object. First this program reads the csv file of incidents (spatial coordinates with time) and transforms it to an integer array. Then both spatial and temporal distances among incidents are calculated for each pair. In the demo, we use 5,000 as the spatial distance band, and 100 as the temporal distance band. Five categories are specified for both space and time. Hence, the five categories for space are: 0, (0, 5,000), (5,000, 10,000), (10,000, 15,000), $\geq$15,000. The first spatial category is the same geographical location and the last one is those beyond 15,000. The five temporal categories are 0, (0, 100), (100, 200), (200, 300), $\geq$300. Combing these possibilities, a 5×5 matrix is formed. The top left corner cell represents the pairs occurring at the same location and at the same time, while the bottom right corner cell locate those beyond the geographical distance of 15,000 and the temporal distance of 300. The traditional space-time interaction occurs among paired events. In other words, we need to assign pairs to its corresponding cells with related spatial and temporal categories. The logic for more than three events follows: in each cell (one of 25 cells in the matrix), check how many three-event groups exist. To be qualified as a three-event group, all the three events must be within the space-time requirement set by that cell for each pair. For example, in the cell of space (0, 5,000) and time (0, 100), if we want to put three-event: a, b, c to this cell, an event must be with such space and time with event b, event b must have such relationship with event c, and event c must have such relationship with event a. The solution is: (1) first run two-event test, get the record file (list all the pairs in the corresponding cell). (2) For each cell, check the existing two-event pairs. For example, in the cell of space (0, 5,000) and time (0, 100), we have (a,b), (a,c), (b,c), (c,d), and (d,e). To be qualified as a three-event, the event must appear at least twice, so a,b,c,d qualifies and e doesn't qualify. Then the search only limits to a, b, c, and d. Eventually we can decide only a,b, and c can be included. This logic applies to all multiple-event series. To illustrate the statistical significance of the observed pair number, the time of incidents are relocated among geographical coordinates in each run of 999 simulations. A loop is used in the program which allows each pair of incidents to be examined one by one. This procedure put the risk of computation time and complexity at the size of incidents, because the intensive-computation part of the program cannot run simultaneously. When the size of events reaches 30,000, paired events calculation with 999 simulations will collapse.

16.3.2 High Performance Solution Over GPU and Keeneland[2]

Emerging advanced computing technologies such as Graphics Processing Units (GPUs), many-core chips such as Intel Many Integrated Core (MIC) architecture, and heterogeneous computer systems such as Keeneland and Blue Waters that combine accelerators and multi-core nodes, can accelerate scientific computation. Today's GPU is high-performance many-core processor. GPUs are designed as general purpose parallel processor with support of accessible programming interfaces and standard programming languages. Impressively as Prof. Jack Dongarra indicated, "GPUs have evolved to the point where many real world applications are easily implemented on them and run significantly faster than on multi-core systems. Future computing architectures will be hybrid systems with parallel-core GPUs working in tandem with multi-core CPUs". Keeneland is a powerful hybrid system sponsored by NSF. At the time of its release in 2010, it ranked 118th in the list of top 500 supercomputers in the world. Keeneland is composed of an HP SL-390 (Ariston) cluster with Intel Westmere hex-core CPUs, NVIDIA Fermi GPUs, and a Qlogic QDR InfiniBand interconnect. The system has 120 nodes with 240 CPUs and 360 GPUs. Each node has 2 Westmere hex-core CPUs, while each CPU has 67 GFLOPS of computing power and three GPUs. Each GPU can generate 515 GFLOPS of computing power. Every four nodes are placed in the HP S6500 Chassis, and every six Chassis is placed in a rack. In total, seven racks are included in the Keeneland system. The Keeneland full-scale system was added to the XSEDE in July 2012.

We have successfully implemented the near-repeat calculation for two event chains over a sample data with 32,507 records on a desktop GPU and the Keeneland, a supercomputer with hybrid architecture that has 240 CPUs and 360 GPUs sponsored by NSF. Through a re-engineering process, the near-repeat calculation is first parallelized on to a NVIDIA GeForce GTX 260 GPU, which has 27 streaming multiprocessors (SM). Each SM has eight CUDA cores as streaming processor (SP). In this GTX 260 with a compute capability of 1.3, up to 1,024 threads can be assigned to each SM. Thus a maximum of $1,024 \times 27 = 27,648$ threads can run in parallel.

Keeneland is a powerful hybrid system jointly developed by Georgia Institute of Technology, the University of Tennessee at Knoxville and the Oak Ridge National Laboratory through sponsorship from NSF. Keeneland is composed of an HP SL-390 (Ariston) cluster with Intel Westmere hex-core CPUs, NVIDIA 6 GB Fermi GPUs, and a QLogic QDR InfiniBand interconnect. The system has 120 nodes with 240 CPUs and 360 GPUs. Each node has 2 Westmere hex-core CPUs, while each CPU has 67 GFLOPS of computing power and three GPUs. Each GPU can generate 515 GFLOPS of computing power. Every four nodes are placed in the HP S6500 Chassis, and every six Chassis is placed in a rack. In total, seven racks are included in the Keeneland system

---

[2] This work was supported partially by the National Science Foundation through the award OCI-1047916.

When the near-repeat calculation is parallelized for calculating two event chains over desktop GPU, while a duplication of the paired values occurs on the GPU side, the duplicated pairs can be eliminated during the integration process on the CPU side. It takes about 48.5 min to complete the entire calculation and simulation processes for a 1,000 runs. Furthermore, the near-repeat calculation was implemented on the Keeneland. Through a combination of MPI and GPU programs, we can dispatch the simulation work onto multiple nodes in Keeneland to accelerate the simulation process. We use 100 GPUs on Keeneland to implement 1,000 simulations. It spends about 264 s to complete this task. If more GPUs were used, the simulation time can be reduced expectedly.

16.4 Conclusion

The importance of space to many socioeconomic theories has been gaining attention and recognition (Krugman 1999; Goodchild et al. 2000; Rey and Ye 2010). The fast growth in socioeconomic dynamics analysis is increasingly seen as attributable to the availability of space-time datasets (Goodchild and Glennon 2008). Research in the spatiotemporally integrated socioeconomic dynamics covers a variety of themes across multiple disciplines coupled with complex methodologies (Stefanidis et al. 2011). However, spatial social scientists have been slower to adopt and implement new spatiotemporally explicit methods of data analysis due to the lack of data and computation power, which becomes a major impediment to promote successful place-based policy implementation and evaluation. The vision and pilot study in this paper aims to develop the computational capability of exploring space-time socioeconomic measurements based on event data, which lend support to the notion that space and time cannot be meaningfully separated. This research bridges emerging advanced computer infrastructure and computing technology with socioeconomic analysis, which is among the burgeoning efforts seeking the cross-fertilization among multiple fast-growing interdisciplinary communities. This research will especially build computing capability to develop, evaluate, and implement a framework to comprehensively quantify the changes and level of hidden variation of space-time event datasets.

This pilot study demonstrates the promising feature of high performance computing on the solution of large-scale space-time interaction, which is a challenging statistical issue in spatiotemporally integrated social science. Researchers are beginning to pay more attention to both spatial and temporal aspects of large amounts of event data. The discovery of repeat and near-repeat phenomena among events is a typical example. A growing list of literature, for instance, has made significant contributions to identifying repeat and near-repeat patterns of crimes. These research findings appeal to police and crime prevention specialists, because of their potential to improve policing policies on detecting the crime hot spots. More specifically, this research provides cues for preventing possible follow-up shootings and can thus be used to direct police activities at finer spatiotemporal scales.

Research on repeat and near-repeat shootings will provide practical information on gun violence prevention. The analysis will directly guide limited police resources to be prioritized in risky locations during risky time periods once a shooting occurs. Nevertheless, to date, little is known about how a series of events are formed at the statistical level due to the lack of such intensive computation design. The statistical limitation of repeat and near-repeat analysis should be noted. For example, this statistical analysis can be considered as a before-and-after analysis of a specific type of event in any specific location and its vicinity. The assumption is that the space-time pattern of such events at any location and its vicinity is random before a shooting occurs. After the shooting, a spatial cluster of arrests is expected near or at the same location. Given that these concentrations of shootings occur in high-crime areas, a series of events, instead of paired events, should be expected (policing activities, followed by shootings, and then by additional police responses). To deal with such time-series spatial data in the context of big data, computation capability is a must. The near repeat test only determines whether there are more event-pairs in a close proximity of space and time than would be expected on the basis of a random distribution. The challenge is that the linkage beyond paired events is ignored in the near-repeat analysis. Space-time chain is defined as a series of more than three incidents which significantly cluster in space and time. However, it is very computational-intensive to do so. Given the example for a sample data with 30,000 events, near repeat calculation and simulation can be a petascale (10^{15}) problem to derive all three event chains, or one run of five event chain calculation is over exascale (10^{18}). With the aid of high performance computing, space-time based preventive patrol and social service towards the most at-risk offenders and places can be much more efficiently implemented by utilizing limited resources.

Through building the capability to explore and compare the potential interactions among emerging themes of event data across different space-time scales and dimensions, this analysis can motivate new queries that are worthy of additional research (Sun et al. 2011). Such data intensive analysis enables access to a much wider thinking which addresses the role of dimensions and scales at different stages of socioeconomic dynamics for more in-depth study. Through exploratory endeavors, this study will motivate social scientists to formulate and verify new hypotheses from theoretical and policy perspectives. This space-time work provides an important contribution to the current spatial science sciences literature, which lacks both real-time data and computation power. The proposed research can also be applied to a wide set of socioeconomic processes with geo-referenced data measured over time. The project will contribute to the promotion of the next generation scientists who have interdisciplinary knowledge and advanced skills for data and computational social science research.

References

Andrienko, N., & Andrienko, G. (2006). Exploratory Analysis of Spatial and Temporal Data: A Systematic Approach. Berlin: Springer

Andrienko, N., & Andrienko, G. (2012). A visual analytics framework for spatio-temporal analysis and modeling. Data Mining and Knowledge Discovery, 1–36. doi:10.1007/s10618-012-0285-7

Anselin, L., From SpaceStat to CyberGIS. (2012). Twenty Years of Spatial Data Analysis Software. *International Regional Science Review, 35*(2),131–157

Batty, M. (2010). The pulse of the city. Environment and Planning B: Planning and Design, 37(4):575–577. doi:10.1068/b3704ed

Bennett, T., & Durie, L. (1999). *Preventing Residential Burglary in Cambridge: From Crime Audits to Targeted Strategies* (Police Research Series Paper 108). London: Home Office

Bernasco, W. (2008). Them again?: Same-offender involvement in repeat and near repeat burglaries. *European Journal of Criminology, 5*, 411–431

Bowers, K. J., & Johnson, S. D. (2004). Who commits near repeats? A test of the boost explanation. *Western Criminology Review, 5*, 12–24

Bowers, K. J., & Johnson, S. D. (2005). Domestic burglary repeats and space-time clusters. *European Journal of Criminology, 2*, 67–92

Brady, T. V. (1996). *Measuring what matters part one: Measures of crime, fear and disorder* (National Institute of Justice: Research in Action Series). Washington: US Department of Justice

Goodchild, M. F., Anselin, L., Appelbaum, R., & Harthorn, B. (2000). Toward spatially integrated social science. *International Regional Science Review, 23*, 139–159

Goodchild, M. F. (2004). GIScience, geography, form, and process. *Annals of the Association of American Geographers 94*(4),709–714

Goodchild, M. F. (2009). Geographic information systems and science: today and tomorrow. *Annals of GIS, 15*(1), 3–9. doi:10.1080/19475680903250715

Goodchild, M. F., & Glennon, A. (2008). Representation and computation of geographic dynamics. In K.S. Hornsby & M. Yuan (Ed.), Understanding Dynamics of Geographic Domains (pp. 13–30). Boca Raton: CRC Press

Grubesic, T. H., & Mack, E. A. (2008). Spatiotemporal interaction of urban crime. *Journal of Quantitative Criminology, 24*, 285–306

Guo, D., & Mennis, J. (2009). Spatial data mining and geographic knowledge discovery-An introduction, Computers. *Environment and Urban Systems, 33*(6), 403–408

Johnson, S. D, & Bowers, K. J. (2004). The burglary as clue to the future: The beginnings of prospective hot-spotting. *European Journal of Criminology, 1*, 237–255

Johnson, S. D., Bernasco, W., Bowers, K. J., Elffers, H., Ratcliffe, J. H., Rengert, G. F., & Townsley, M. (2007). Space-time patterns of risk: A cross national assessment of residential burglary victimization. *Journal of Quantitative Criminology, 23*, 201–219

Johnson, S. D., Summers, L., & Pease, K. (2009). Offender as forager? A direct test of the boost account of victimization. *Journal of Quantitative Criminology, 25*, 181–200

Knox, G. (1963). Detection of low intensity epidemicity: Application to cleft lip and palate. *British Journal of Preventive and Social Medicine, 17*, 121–27

Knox, G. (1964). Epidemiology of childhood leukaemia in Northumberland and Durham. *British Journal of Preventive and Social Medicine, 18*, 17–24

Krugman, P. (1999). The role of geography in development. *International Regional Science Review, 22*(2), 142–161

Maguire, E. R., Willis, J. A., Snipes, J. B., & Gantley, M. (2008). Spatial concentrations of violence in Trinidad and Tobago. *Caribbean Journal of Criminology and Public Safety, 13*, 48–92

Morgan, F. (2001). Repeat burglary in a Perth suburb: Indicator of short-term or long-term risk? In G. Farrell, & K. Pease (Ed.), *Repeat Victimization* (pp. 83–118). Monsey, New York: Criminal Justice Press

Ratcliffe, J. H., & Rengert, G. F. (2008). Near-repeat patterns in Philadelphia shootings. *Security Journal, 21*, 58–76

Rey, S. J. & Ye, X. (2010). Comparative spatial dynamics of regional systems. In Pàez, A., Gallo, J. L., Buliung, R., & Dall'Erba, S. (Ed.), Progress in Spatial Analysis: Methods and Applications (pp.441–463). London, New York: Springer

Sagovsky, A., & Johnson, S. D. (2007). When does repeat burglary victimization occur? *The Australian and New Zealand Journal of Criminology, 40*, 1–26

Stefanidis, A., Crooks, A., & Radzikowski, J. (2011). Harvesting ambient geospatial information from social media feeds. GeoJournal. doi:10.1007/s10708-011-9438-2

Sun, A., Valentino-DeVries, J., & Seward, Z. (2011). A week on Foursquare. The Wall Street Journal. Available online at: http://graphicsweb.wsj.com/documents/FOURSQUAREWEEK 1104/ [Last Accessed 11/12/2011]

Tobler, W. R. (1970). A Computer Movie Simulating Urban Growth in the Detroit Region. *Economic Geography, 46*, 234–240. doi: 10.2307/143141

Townsley, M., Homel, R., & Chaseling, J. (2000). Repeat burglary victimization: Spatial and temporal patterns. *Australian and New Zealand Journal of Criminology, 33*, 37–63

Trickett, A., Osborn, D. R., Seymour, J., & Pease, K. (1992). What is different about high crime areas? *British Journal of Criminology, 32*, 81–89

Warf, B., & Sui, D. (2010). From GIS to neogeography: ontological implications and theories of truth. *Annals of GIS, 16*, 197–209

Wells, W., Wu, L., & Ye, X. (2012). Patterns of near-repeat gun assaults in Houston. *Journal of Research in Crime and Delinquency, 49*, 186–212

White House (2012). Executive Office of the President (March 2012). "Big Data Across the Federal Government". White House. http://www.whitehouse.gov/sites/default/files/microsites/ostp/big_data_fact_sheet_final.pdf (Last Access on: 01-29-2013)

Ye, X., & Carroll, M. (2011a). Exploratory space-time analysis of local economic development. *Applied Geography, 31*, 1049–1058

Ye, X., & Carroll, M. (2011b). Warn notice toolbox: open-source geovisualization of large lay-off events, GeoInformatics 2011 proceedings DOI: 10.1109/GeoInformatics.2011.5981136

Ye, X., & Liu, L. (2012). Special issue on Spatial crime analysis and modeling, *Annals of GIS, 18*(3), 157–241

Ye, X., & Rey, S. J. (2011). A framework for exploratory space-time analysis of economic data. Annals of Regional Science. DOI: 10.1007/s00168-011-0470-4

Chapter 17
Opportunities and Challenges for Urban Land-Use Change Modeling Using High-Performance Computing

Qingfeng Guan and Xuan Shi

Abstract Simulating urban land-use changes involves both high modeling and computational complexities. This paper focuses on a typical spatio-temporal modeling method that has been commonly used in urban land-use change studies—Cellular Automata (CA). After reviewing the recent development of utilizing various parallel computing technologies (e.g., computer clusters and Graphics Processing Unit [GPU]) in CA-based urban models, this paper presents a pilot study, in which a classical CA model, the Game of Life, was implemented as a parallel program over the GPU/CPU heterogeneous cluster architecture, and 300+ speed-up was achieved using 20 GPUs. In conclusion, emerging high-performance computing technologies, such as GPU/CPU heterogeneous cluster architecture, provide promising potentials to overcome the computing obstacle of urban land-use change models, and enable researchers to examine, validate and advance urban land-use change theories and derive sound urban planning strategies. To efficiently utilize the computing power of the GPU/CPU clusters, hybrid parallelism must be implemented to coordinate the computing among GPU/CPU nodes, as well as among the threads on each GPU. However, implementing such hybrid parallelism is challenging for its high development complexity.

Keywords Parallel computing • GPU • Heterogeneous cluster architecture • Urban • Land-use change

Q. Guan (✉)
Faculty of Information Engineering, China University of Geosciences (Wuhan),
Wuhan, Hubei 430074, China
e-mail: guanqf@cug.edu.cn

X. Shi
Department of Geosciences, University of Arkansas, Fayetteville, AR 72701, USA
e-mail: xuanshi@uark.edu

X. Shi et al. (eds.), *Modern Accelerator Technologies for Geographic Information Science*,   227
DOI 10.1007/978-1-4614-8745-6_17, © Springer Science+Business Media New York 2013

17.1 Introduction

Simulating urban land-use changes (LUC) has been a challenging task because of the spatiotemporal complexities of interrelationships and interactions between the urban land system and related natural/socioeconomic systems. The difficulty of modeling urban LUC can be aggravated by the massive computational intensity caused by complicated algorithms and large datasets that are often required in the simulation. Some large scale simulations have been infeasible because they are computationally intractable using conventional desktop computers.

In order to reduce both the modeling and computational complexities in spatiotemporal simulations, researchers often had to make subjective and/or simplifying assumptions. However, such simplifying approaches had raised some serious scientific questions in regard to the validity and soundness of the findings resulted from these models, because whether these assumptions could generate reliable calibration and simulation results and lead to unbiased and accurate scientific conclusions has not been sufficiently studied yet. To investigate the potential problems and advance our understanding and theories of urban land dynamics, we must devise approaches to reduce, or even eliminate, these assumptions.

Recent advancements in high-performance computing (HPC) infrastructure provide potential solutions to the above problems. Emerging advanced computing technologies, such as Graphics Processing Units (GPUs) and heterogeneous cluster computing systems that combine multiple GPU accelerators and Central Processing Units (CPUs), have been significantly improving the performance of scientific computation in a variety of domains. Therefore, it is time for geographers and geospatial scientists to examine, validate and advance urban LUC theories as the technological solutions and computing infrastructure are increasingly mature and efficient for such kind of investigations.

17.2 Spatiotemporal Modeling of Urban Land-Use Changes

Many approaches exist to model urban land-use changes and associated natural and socio-economic dynamics. A large proportion of them are based on variants of the Cellular Automata (CA) model, a discrete computational model used to simulate dynamic spatial processes through a set of transition rules. A classical CA model has a set of identical elements, called cells. Each cell is located in a regular, discrete space, called a cellspace. Each cell is associated with a state within a finite set. The model evolves in discrete time steps, changing the states of all its cells according to transition rules, homogeneously and synchronously applied at every step. The new state of a certain cell depends on the previous states of the cells within its neighborhood. CA models have been widely used in geographic research to simulate complex spatiotemporal phenomena, including land-use and land-cover change (Batty et al. 1999; Couclelis 1997; Wu and Webster 1998; Li and Yeh 2000; Liu and Phinn 2003), wildfire propagation (Clarke et al. 1995), and freeway traffic (Nagel and Schreckenberg 1992; Benjamin et al. 1996).

A typical example is the SLEUTH model, one of the most widely used urban LUC models (Clarke et al. 1997; Clarke and Gaydos 1998; Silva and Clarke 2002). The core of SLEUTH is an urban growth model, which uses a modified CA to simulate the spread of urbanization across a landscape. The behavior of the simulation is determined by five parameters (also termed coefficients), each ranging from 0 to 100. Four growth rules are applied in sequence on the space during each growth cycle, which represents a year of urban growth.[1]

Calibration is needed to determine the appropriate parameter values so that SLEUTH can produce realistic simulation results. The basic calibration procedure of SLEUTH uses the brute-force method, which statistically compares multiple test results produced using combinations of parameter values with the real historical dataset, in order to determine the best-fit parameter combination(s). In addition, to simulate the random processes during urban growth, the Monte Carlo method is applied multiple times, and the outcomes are stored as the cumulative probabilities of change over multiple runs. In practice, 10–100 Monte Carlo iterations for each parameter combination are suggested, although fewer may be better than more (Goldstein et al. 2005).

All of the above together make the calibration highly computationally intensive. A 12-year (1986–1998) simulation over a small-sized dataset (2,074×486) of Santa Barbara County in California took only 1 seconds to complete on a desktop PC. However, a comprehensive calibration over the same dataset and time period to examine all 101^5 parameter combinations with only 1 Monte Carlo iteration was estimated to take over 300 years to complete. This places the SLEUTH model at the edge of computational tractability. The current version of SLEUTH model uses a simplifying assumption to ignore those "unimportant" parameter values during seeking the best-fit combination(s), which is that the parameters affect the simulation results in a linear manner. However, due to the random processes involved in the transition rules, the relationships between the parameters/factors and LUC simulations are very likely non-linear. Thus the calibration results based on such simplifying assumptions are hardly fact-proven (due to the incomprehensive calibration), less reliable, and may lead to inaccurate scientific conclusions and improper land management decisions (Dietzel and Clarke 2007).

Alternatively, researchers have used Computational Intelligence (CI) methods to either seek the best-fit parameter combination(s) without evaluating all the combinations, or construct transition rules for the model (see for example Li and Yeh 2002; Wu and Silva 2010; Liu et al. 2010; Li et al. 2013). However, the computational burden of CA itself is not diminished by CI methods, and the computational intensity may still exceed the capacity of a desktop computer when using complex transition rules and massive datasets.

---

[1] For details, see http://www.ncgia.ucsb.edu/projects/gig/About/gwRules.htm.

17.3   Exploratory Studies on High-Performance Spatial CA

17.3.1   Parallel Spatial CA

The classical CA model has been recognized to be a natural parallel computing system as the transition rules are applied to the cells homogeneously and synchronously in parallel (Bandini et al. 2001). The cellspace can be easily decomposed into a set of small sub-cellspaces and assigned onto multiple computing units (e.g., CPUs and CPU cores) to be processed simultaneously. Several general parallel CA-based simulation systems have been developed. Examples include the Cellular Automata environment for systEms ModeLing (CAMEL) and CellulAR Programming EnvironmenT (CARPET) language (Spezzano and Talia 1999), and Cell Driver, a CA modeling module of NEMO (Hecker et al. 1999). Both CAMEL and Cell Driver were built based on the Message Passing Interface (MPI), a generic parallel programming library that is available on most parallel computing systems.

Guan and Clarke (2010) developed an open-source general-purpose parallel Raster Processing programming Library (pRPL), for non-specialist scientists to easily parallelize their own raster processing algorithms. pRPL supports multi-layer algorithms that are commonly used in geospatial applications, including spatial CA. pRPL provides multiple data decomposition methods, including a spatially-adaptive quad-tree-based (QTB) decomposition method for situations when the computational intensity is extremely heterogeneous over space. pRPL also automatically takes care of some complicated processes that are required in parallel computing, e.g., communication, synchronization and load-balancing, thus provides transparent parallelism for users. A parallel urban LUC model, pSLEUTH, was developed based on the SLEUTH model using pRPL. Experiments with real-world datasets showed that pSLEUTH greatly reduced the computing time for the calibration process, achieving a speed-up of 24 using 32 CPU cores on a computer cluster composed of 128 dual CPU 3.06 GHz Xeon nodes with 2 GB RAM each.

However, all above parallel CA systems are based on conventional CPU-only parallel computing architectures such as multi-core CPUs and computer clusters. Large-scale parallel computing facilities are extremely expensive and require tremendous amount of financial and labor investments, and very limited to public access. Also, the waiting time in a job queue on a computer cluster may exceed the actual computing time, which makes the performance gain from parallel computing less meaningful. An emerging accelerator technology, GPU with the Compute Unified Device Architecture (CUDA), is able to accelerate the computation processes by deploying hundreds of computing cores on the GPU with very low costs. A PC equipped with a GPU is considerably cheaper than a computer cluster that has the same number of cores. GPUs are very suitable for parallel matrix manipulation and processing, which is similar to CA computing. Some efforts have been made to implement CA models on GPUs, which generated high speed-ups (Thor 2008; Li et al. 2012). Moreover, the heterogeneous computer cluster architecture can generate massive computing power by coordinating a set of computational nodes that consists of one or more CPU(s) and GPU(s). The heterogeneous cluster architecture has been adopted to build high-end computing platforms to handle super large-scale scientific and engineering computation.

17.3.2   Accelerating CA on GPUs and Heterogeneous Computer Systems

In order to explore the possibility and validity of utilizing the emerging HPC technologies in urban LUC studies, we have successfully prototyped parallel CA models on both GPU-equipped PCs and GPU/CPU heterogeneous clusters. The Game of Life (GOL) is a well-known classical CA model. Based on the transition rule, a cell can live or die depending on the condition of its 3×3 neighborhood. As a result, the living status of the cells can represent various spatial patterns throughout the course of iterations. The pseudo code of the GOL's transition rule is as follows:

```
FUNCTION Transition (cell, time_t)
  n = number of alive neighbors of cell at time_t
  IF cell is alive at time_t
    IF n ≥ 4
    THEN cell dies of overcrowding at time_t+1
    IF n ≤ 1
    THEN cell dies of loneliness at time_t+1
    IF n = 2 OR n = 3
    THEN cell survives at time_t+1
  ELSE (i.e., cell is dead at time_t)
    IF n = 3
    THEN cell becomes alive (i.e., born) at time_t+1
```

In Table 17.1, the leftmost figure displays the initial status for a 10,000 by 10,000 matrix in which half of the matrix would contain living cells. After 100 iterations, many cells may die and the right-most figure displays the result of the simulation.

Here we introduce the steps taken to create efficient parallel implementations of GOL. In order to ensure that all solutions generate the same result, we create a matrix file that contains the initial living status of randomly generated cells. All versions of the program share a similar initialization phase where this matrix file is read into the appropriate array or arrays in the case of the MPI/CUDA program. Each of the programs was benchmarked against the same set of matrices for 100 iterations. We tested all solutions using the matrix that has a dimension of $10,000 \times 10,000$, which was initially seeded with half of them as living and half dead.

The GOL was first implemented in a serial C program. A 100-iteration simulation over a $10,000 \times 10,000$ cellspace was accomplished in about 100 minutes on a desktop PC with a 1.60 GHz dual-core CPU. Within the serial C program, for each iteration, each cell will change its living status by examining the living status of its neighbors. Finally the number of living cells is accumulated.

Table 17.1 $t=0$ [left] through $t=99$ [right]

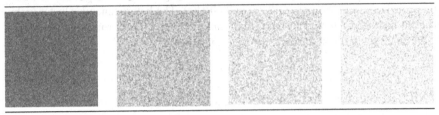

The GOL was then parallelized into a CUDA program, called GPU-GOL. The GOL's transition rule was implemented as a kernel function. During the simulation, a large number of computational threads are simultaneously invoked on the GPU, each executing an instance of the kernel and applying the transition rule on a small proportion of the whole cellspace.

Since counting the living cells is a sequential process, it was first excluded from the GPU kernel program that implements the transition rules of GOL, while counting the neighbors is a device function. After the result is copied back from device to host, the total number of living cells will be counted in sequential process. The GPU-GOL experiments were conducted on a desktop PC with a NVIDIA GeForce GTX 260 GPU, which has 27 streaming multiprocessors (SM), and is able to run up to 27,648 threads in parallel. The simulation at a size of 10,000×10,000 for 100 iterations took about 6 minutes to complete, achieving a speed-up of 16.7.

Further improvement was taken to implement the process of counting the number of living cells through atomicAdd function within the kernel program. GOL simulation at a size of 10,000×10,000 for 100 iterations can be completed in about 22 seconds on a single Tesla C2075 Fermi GPU or a single Tesla K20 Kepler GPU, achieving a speedup of 13 in comparison to GTX 260. When shared memory was utilized, better performance could be achieved even over a single GPU, though further examination needs to be conducted to validate the solution over different platforms.

Since a single GPU may not efficiently handle the scalability of computation due to the memory limit on individual GPU, we would like to explore the potential of utilizing multiple GPUs to resolve this problem. Keeneland's hybrid architecture exemplifies its superiority in manipulating the large scale cellular automation computation like GOL. Keeneland is composed of an HP SL-390 (Ariston) cluster with Intel Westmere hex-core CPUs, NVIDIA 6GB Fermi GPUs, and a Qlogic QDR InfiniBand interconnect. The system has 120 nodes, each with two CPUs and three GPUs, while all CPUs and GPUs are bridged together through one I/O hub from which the CPUs can read/write data.

To efficiently utilize and manage the GPU resources in Keeneland, we implemented a combination of MPI and CUDA programs to parallelize the GOL computation on 20 GPUs. Although the CUDA kernel for this implementation is nearly identical to what is implemented in the single-GPU program, data communication become a serious problem due to the strong dependency between the data segments distributed onto different GPU processors.

A row-based data partitioning approach was applied to distribute data segments onto multiple GPUs. We tried to decompose the entire matrix into multiple sections based on the number of GPUs we utilized. In this case, each MPI process reads in a unique portion of the matrix file based on process rank. When the original matrix is split in this way and updated separately on different GPUs, each GPU needs to obtain extra rows of information hosted by the other processors because the state of the cells along the matrix boundaries are dependent upon cells which now are in other sub-matrices handled by different MPI processes.

In order to exchange these boundary rows between the neighboring processors, we applied the SEND and RECV functions in MPI for sending and receiving the boundary rows (i.e. head and tail for each block of the grid) between neighboring processors ranked in MPI, and then copying these rows to the GPU memory. Script 17.1 describes how to handle the data transfer between the host CPU and the GPU, and coordinate the computational threads on the GPU. For each iteration, SEND and RECV functions are first implemented to construct the local data segments to be calculated on each node. The GPU on each node executes the kernel function (i.e., the transition rule) covering one portion of the matrix.

```
for(k = 0; k <ITERATION ; k++) {
if (myrank % 2 == 1) {
    // send tail and receive head
    MPI_Send(...);
    MPI_Recv(...);
    // send head and receive tail
    MPI_Send(...);
    MPI_Recv(...);
}
else {
    // receive head and send tail
    MPI_Recv(...);
    MPI_Send(...);
    // receive tail and send head
    MPI_Recv(...);
    MPI_Send(...);
}
```

Script 17.1 Implementing SEND/RECV for data exchange in MPI program

When 20 GPUs on Keeneland were used, a 100-iteration GOL with a size of $10,000 \times 10,000$ was completed in 20 seconds. The results were the same as what was generated by the serial C program and GPU-GOL. In short, the computing time was significantly reduced from 100 minutes to 20 seconds, achieving a speed-up of 300. When the atomicAdd approach was applied, GOL over the same size of matrix can be accomplished in about 2 seconds when 20 GPUs were used.

The parallel solution over heterogeneous computer architecture and systems have shown promising prospect to break through the computational bottleneck of CA models that include complex transition rules and use massive datasets. By simply changing the CA's transition rules to simulate more complex spatiotemporal processes, we may use such an approach to conduct some large-scale urban LUC simulations within a practical length of time.

17.4 Conclusion

Modeling the spatiotemporal dynamics of land use and land cover in the urbanization process often involves complex algorithms and large volume of datasets, which greatly increases the computational intensity, hence sometimes requires unfeasibly long computing time. Simplifying assumptions have been used in previous studies to reduce the computational intensity, but they may generate unreliable results and lead to inaccurate scientific conclusions and improper land management decisions. Emerging high-performance computing technologies, such as GPU and GPU/CPU heterogeneous cluster architecture, provide promising potentials to overcome the computing burden of urban LUC models, thus to enable researchers to examine, validate and advance urban LUC theories and derive sound urban planning strategies. To efficiently utilize the computing power of the GPU/CPU heterogeneous clusters, hybrid parallelism must be implemented to coordinate the parallel computing among GPU/CPU nodes, as well as among the threads on each GPU. However, implementing such hybrid parallelism is challenging for its high development complexity in integrating MPI and CUDA.

In this pilot study, we demonstrated the potential for accelerating CA applications using parallel implementation on hybrid computer clusters. While parallel implementation of CA through MPI + GPU has achieved significant performance improvement, the emerging new architecture of Intel's Many-Integrated Core (MIC) could be another potential accelerator technology for urban LUC simulations. It was found from our other initiatives that the simple MPI-direct-host programming model on Intel MIC cluster can achieve a performance equivalent to the MPI + GPU model on GPU clusters when the same number of processors are allocated for Kriging interpolation calculation and for unsupervised image classification.

Exploring efficient cross-node communication mechanism could be a key component in the future work so as to achieve a strong scalability for CA-based applications running on multiple parallel nodes. For example, the latest Tesla K20 Kepler GPU is able to outperform the Fermi GPU for most applications without special performance tuning. However, K20's direct cross-GPU communication mechanism needs to be explored and deployed to enhance CA-based modeling that has intensive data communication between the nodes. Meanwhile solutions based on Intel MIC architecture is worthy to try since each MIC core has direct support of MPI, making it straightforward to port MPI + CPU code to MIC cluster to achieve significant performance improvement. Exploring a combination of MPI and OpenMP solutions will help handle inter-node and intra-node communications to efficiently utilize the heterogeneous computer architecture and systems.

Acknowledgements This research was supported partially by the National Science Foundation through the award OCI-1047916.

References

Bandini, S., Mauri, G. & Serra, R., 2001. Cellular Automata: From a Theoretical Parallel Computational Model to Its Application to Complex Systems. *Parallel Computing*, 27, pp. 539–553.

Batty, M., Xie, Y. & Sun, Z., 1999. Modeling urban dynamics through GIS-based cellular automata. *Computers, Environment and Urban Systems*, 23(3), pp.205–233.

Benjamin, S.C., Johnson, N.F. & Hui, P.M., 1996. Cellular automata models of traffic flow along a highway containing a junction. *Journal of Physics A: Mathematical and General*, 29(12), pp.3119–3127.

Clarke, Keith C. & Gaydos, L.J., 1998. Loose-coupling a Cellular Automaton Model and GIS: Long-term Urban Growth Prediction for San Francisco and Washington/Baltimore. *International Journal of Geographical Information Science*, 12(7), pp.699–714.

Clarke, Keith C., Hoppen, S. & Gaydos, L., 1997. A Self-modifying Cellular Automaton Model of Historical Urbanization in the San Francisco Bay Area. *Environment and Planning B: Planning and Design*, 24(2), pp.247–261.

Clarke, Keith C., Riggan, P. & Brass, J.A., 1995. A cellular automaton model of wildfire propagation and extinction. *Photogrammetric Engineering and Remote Sensing*, 60(11), pp.1355–1367.

Couclelis, H., 1997. From Cellular Automata to Urban Models: New Principles for Model development and implementation. *Environment and Planning B: Planning and Design*, 24(2), pp.165–174.

Dietzel, Charles & Clarke, Keith C, 2007. Toward Optimal Calibration of the SLEUTH Land Use Change Model. *Transactions in GIS*, 11(1), pp.29–45.

Goldstein, N.C., Dietzel, C. & Clarke, K. C., 2005. Don't stop 'til you get enough–sensitivity testing of Monte Carlo iterations for model calibration. In *Proceedings of the 8th International Conference on GeoComputation*. Ann Arbor Michigan.

Guan, Q. & Clarke, K. C., 2010. A general-purpose parallel raster processing programming library test application using a geographic cellular automata model. *International Journal of Geographical Information Science*, 24(5), pp.695–722.

Hecker, C. et al., 1999. System Development for Parallel Cellular Automata and Its Applications. *Future Generation Computing Systems*, 16(2–3), pp.235–247.

Li, D. et al., 2012. GPU-CA model for large-scale land-use change simulation. *Chinese Science Bulletin*, 57(19), pp.2442–2452.

Li, X. et al., 2013. Calibrating cellular automata based on landscape metrics by using genetic algorithms. *International Journal of Geographical Information Science*, 27(3), pp.594–613.

Li, X. & Yeh, A.G.O., 2000. Modelling Sustainable Urban Development by the Integration of Constrained Cellular Automata and GIS. *International Journal of Geographical Information Science*, 14(2), pp.131–152.

Li, X. & Yeh, A.G.O., 2002. Neural-network-based Cellular Automata for Simulating Multiple Land Use Changes Using GIS. *International Journal of Geographical Information Science*, 16(4), pp.323–343.

Liu, Xiaoping et al., 2010. Simulating land-use dynamics under planning policies by integrating artificial immune systems with cellular automata. *International Journal of Geographical Information Science*, 24, pp.783–802.

Liu, Y. & Phinn, S.R., 2003. Modelling urban development with cellular automata incorporating fuzzy-set approaches. *Computers, Environment and Urban Systems*, 27(6), pp.637–658.

Nagel, K. & Schreckenberg, M., 1992. A cellular automaton model for freeway traffic. *Journal of Physics I France*, 2, pp.2221–2229.

Silva, E.A. & Clarke, Keith C., 2002. Calibration of the SLEUTH Urban Growth Model for Lisbon and Porto. *Computers, Environment and Urban Systems*, 26(6), pp.525–552.

Spezzano, Giandomenico & Talia, Domenico, 1999. Programming Cellular Automata Algorithms on Parallel Computers. *Future Generation Computing Systems*, 16(2), pp.203–216.

Thor, M., 2008. *Performance comparison of CPU and GPU based simulation of an avalanche using a cellular automata*. Master Thesis. Sweden: Ume°a University.

Wu, F. & Webster, C.J., 1998. Simulation of Land Development through the Integration of Cellular Automata and Multi-criteria Evaluation. *Environment and Planning B*, 25(1), pp.103–126.

Wu, N. & Silva, E.A., 2010. Artificial Intelligence Solutions for Urban Land Dynamics: A Review. *Journal of Planning Literature*, 24(3), pp.246–265.

Chapter 18
Modern Accelerator Technologies for Spatially-Explicit Integrated Environmental Modeling

Dali Wang and Shujiang Kang

Abstract Integrated environmental system modeling is a promising practice to better understand interactions between the human and major components of environmental systems. Spatially-explicit modeling approach is a natural choice for those modeling activities because landscape is one of few shared components across all models. In this chapter, authors first present several examples of integrated environmental modeling, then explain the importance and role of spatially-explicit landscape in those efforts. After that, authors review current applications of modern accelerator technologies in environmental modeling. At last, authors identify the several potential research of integrated environmental system model, which could further be developed in the near future using modern accelerator technologies.

Keywords Integrated environmental modeling • Spatially-explicit modeling • accelerator technologies • Parallel computing

18.1 Introduction

Environmental system modeling presents a variety of challenges. Through the past several decades, along with the rapid development of computing technologies, and strong interests of understanding large-scale environmental phenomena, many computer models have been developed to capture our knowledge on numerous facts of environmental systems, and to explore better options for system-wide management (AERMIC 2004; ATHENA 2007; EPANET 2008; eQUEST 2010). The complexity of

D. Wang (✉) • S. Kang
Oak Ridge National Laboratory, Oak Ridge, TN 37831, USA
e-mail: wangd@ornl.gov; kangs@ornl.gov

X. Shi et al. (eds.), *Modern Accelerator Technologies for Geographic Information Science*, 237
DOI 10.1007/978-1-4614-8745-6_18, © Springer Science+Business Media New York 2013

environmental models varies greatly and represents modelers' own view on the environmental system. For example, many models were developed, focusing on a specific aspect or component of an environmental system (Fernando et al. 2001; Makropoulos et al. 2008; Trusty 2004; Waddell 2000). Researchers have emphasized integrated environmental system modeling (Chen et al. 2006; Li et al. 2007) on advanced computing platforms, which involves complex interactions between some or all components of an integrated environmental system, and results in model systems that link multiple components. Using those model systems, researchers can investigate the system behaviors using several modeling approaches developed by different research groups. There are non-disputable advantages to embrace this modeling approach. It is a very practical method to leverage existing modeling efforts (especially based on open source models) into a comprehensive model system, which otherwise cannot be built by any single institution (Voinov et al. 2008). Those model systems can naturally broaden community engagement. Also from the system design perspective, multi-component system will impose restrictions over each individual component and provide a way to valid those subsystems (Li et al. 2007).

Herein, considering the general interest of readers of this book, authors focus on applications of simulation framework for multi-component environmental systems over spatially-explicit landscape. This chapter is organized as following: authors first present some examples of spatially-explicit, integrated environmental system modeling. Then authors explain their opinions that modern accelerator technologies (with focus on Graphic Processing Unit (GPU) and multicore) are much needed to address some aspects of those computational challenges. At last, authors identify several potential research areas that could take advantage of modern accelerator technologies to meet computational challenges associated with integrated environmental modeling over landscape.

18.2 Three Examples of Spatially-Explicit Integrated Environmental Modeling (SIEM)

This section reviews several examples of integrated environmental modeling efforts. They are (1) ecosystem modeling over landscape, (2) sustainable urban infrastructure, (3) spatially-explicit agroecosystem modeling.

18.2.1 Ecosystem Modeling over Landscape

From a mathematical or computational ecology perspective, ecosystem modeling presents a different set of challenges. Much of classical ecological theory originates from very simple differential equations in which a single variable represents population densities. Although highly influential in ecological theory, the models' aggregated form is particularly difficult to relate to observational biology. Applying them to complex natural systems with spatially and temporally varying environmental factors, for example, typically produces analytically intractable models that must be investigated numerically. Researchers have begun emphasizing integrated, multicomponent ecosystem models (Gross and DeAngelis 2002). These models adapt a linear food chain conceptual model, involve complex interactions between some or all of an ecosystem's trophic layers (feeding levels), resulting in models that link multiple components that researchers can model using several different mathematical approaches for organisms at different trophic positions. Specifically, at a lower trophic level, compartment models can be established to place more emphasis on the kinetics of the ecosystem's energy or nutrient flow, or the pollutant transport within the food web. At a higher level—especially for endangered species with small populations—researchers could use individual based models to monitor and simulate each individual member's basic behaviors, and to indicate the global consequences of local interactions among population members. In the middle trophic level, Structured models average certain population characteristics and attempt to simulate changes in these characteristics for the whole population.

Across Trophic Level System Simulation (ATLSS), developed at the University of Tennessee, is an example of spatially-explicit ecosystem modeling system. ATLSS is designed to assess how alternative water management plans for regulating water flow across the Florida Everglades' landscape will affect key biota. ATLSS's immediate objective is to assist various stakeholders in assessing the biotic impacts of alternative future scenarios for restoring South Florida's natural systems. The long-term goals are to better understand how the biotic communities of South Florida are linked to various physical driving influences—particularly hydrology—and to provide a predictive tool for both scientific research and ecosystem management (ATLSS). Within ATLSS, compartment models deal with variables representing spatially localized biota—mainly the biomasses of lower trophic-level organisms, such as algae, which only interact locally. The age- and size-structured population and community models represent intermediate trophic levels, such as fish (Gaff et al. 2000; Wang et al. 2006), macroinvertebrates, and small nonflying vertebrates. Finally, individual-based models represent populations of top predators and other large-bodied species, such as wading birds (Wolf 1994) and panthers (Comiskey et al. 1994). The internal connections between all the models is developed based on a spatially-explicit landscape library (Duke-Sylvester and Gross 2002). A high performance multilayered software system was developed, which includes a data assimilation layer, a components layer for ecological models, and an information analysis and data representation layer (Wang et al. 2005a). This integrated, spatially explicit ecosystem modeling requires tightly coupled modeling framework (Wang et al. 2007, 2011) and parallel computations (Wang et al. 2008, 2006) on a variety of computing infrastructure. Furthermore, grid computing has been adapted to deliver advanced ecosystem simulation functionalities to naive computer users (biologists, natural resource managers, and so on) (Wang et al. 2005a,b).

18.2.2 Sustainable Urban Infrastructure Modeling

Dense urban areas have complex and interdependent infrastructure systems. With the urban infrastructure expected to increase by more than 40 % between 2000 and 2030 (Nelson 2004), an opportunity exists to significantly change how urban communities consume energy/resources and produce wastes. The current generation of urban infrastructure is predicated on the abundance of cheap carbon-based fuels and plentiful non-renewable resources, which is not sustainable. A Sustainable Urban Infrastructure (SUI) system is being developed at the Georgia Institute of Technology to provide decision support tools for the next generation of urban infrastructure as well as for the renovation of the current urban infrastructure, to use renewable energy/resources and to meet societal, economic and environmental goals. Those decision support tools will allow stakeholders to quantitatively evaluate the benefits of new technologies and to produce a blueprint for sustainable and resilient urban infrastructures as well as the cyberinfrastructure required to monitor them. Technically, SUI will coordinate and leverage experiences from existing projects: (1) examining the resiliency and sustainability of distributed water, energy, and recycling systems; (2) understanding hazard vulnerability and simulating flood protection and river restoration measures; (3) examining the durability and recyclability of materials that are used in urban infrastructure; (4) understanding market forces that shape the demand for more sustainable urban infrastructure; (5) forecasting urban growth and redevelopment; (6) designing high-performance buildings; (7) developing sensors for monitoring urban infrastructure and material flows; (8) predicting economic and social losses due to infrastructure destruction, including understanding responses to changes in urban infrastructure; (9) simulating earthquake and hurricane impacts on infrastructure; (10) understanding future climate impacts on urban areas; (11) creating carbon mitigation strategies for cities; and (12) developing a blueprint for more sustainable urban infrastructure. From software design perspective, based on spatially-explicit landscape, a flexible multi-component modeling framework has being developed to integrate an urban growth/land use forecasting model with the following infrastructure and natural system models: (1) building stock, (2) air quality, (3) transportation, (4) climate/heat island, (5) water supply, (6) wastewater, (7) water quality, (8) stormwater runoff, and (9) electric power. The SUI modeling framework will embed critical engineered infrastructure systems within their social, economic and natural environment. The framework will support the modeling of interactions between and among infrastructure systems and with the relevant social, economic and natural environmental systems at different temporal and spatial scales. Specifically, the high performance multi-component software system combines several existing models, such as UrbanSim (Waddell 2000), TransSims (Barrett et al. 1995), Community Multiscale Air Quality (CMAQ) model (Byun and Schere 2006), PECAS (Hunt and Abraham 2005), Xplorah (Delden et al. 2008) and What If? (Klosterman 1999), where available, and will develop new models as required to fill important gaps in our understanding of infrastructure system performance and interactions under stress.

18.2.3   Spatially-Explicit Agroecosystem Modeling

Multiple concerns over the impact of wide scale changes in agricultural ecosystems have motivated comprehensive analysis of environmental sustainability of food and biofuel production (Lobel et al. 2011). These call for spatial-explicit high-resolution land management models that enable comprehensive analysis of agricultural natural resources for decision-making. Agroecosystem modeling applications have expanded beyond their traditional role in summarizing data and interpreting information from field trials. They have been used to provide information for guiding regional and national scale economic analysis (Adams et al. 1996; Edmonds et al. 1997), and to estimate regional-scale environmental services and impacts (Thomson et al. 2005) and global change impacts on agricultural production (VEMAP_Members 1995). For agroecosystem simulation models to perform well, precise inputs of edaphic conditions, management operations, and weather forcing are required. Recent developments in soil resource mapping, land-use remote sensing, availability of inventory data like crop yield, and consistent fine spatial resolution weather data products enable modeling exercises to meet the demands. The increase of the data volumes and the computational demands for conducting the large number of simulations requires a framework that incorporates these features into a high performance computing environment, along with tools for simulation input data preparation, and for the simulation output data visualization and analysis over spatially-explicit landscape. A typical example of these efforts is the high-performance computational framework for spatially-explicit agroecosystem modeling (Wang et al. in press). Based on spatially-explicit landscape concepts, this framework consists of four key steps: (1) landscape-based simulation data preparation, (2) site-based agroecosystem simulation on high performance computers, (3) cross-landscape data management, and (4) exploratory data analysis over landscape. The batched simulations were executed using a high performance computing environment (HPC-EPIC), which was designed on basis of a widely applied agroecosystem model, Environmental Policy and Impacted Climates (EPIC) at Oak Ridge National Laboratory (Nichols et al. 2011). HPC-EPIC provides an efficient computational approach for deployment of agricultural management at high-resolution spatial scales over large scape spatially-explicit landscape (Kang et al. 2012). The HPC-EPIC has been applied to production and environmental analysis of regional and global ecosystem. In the regional simulation designed for four counties at 56-m resolution, 140,000 simulations were finished in a total of 10 h on an HPC cluster using 20 nodes. Totally, 62,482 simulation units were organized for a global analysis of bioenergy biomass production analysis, and simulations were completed within 3 h. However, since the original agroecosystem model was designed at fine resolution (at field plot level), therefore, a huge number of simulations are needed to cover large geographic regions, even after using the Simulation Unit (SU) concept. Each agroecosystem model simulation produces very detailed agricultural (such as crop yield, biomass, etc.) and environmental information (such as soil organic carbon and nitrogen, and soil water, etc.). Analysis and management of the large amount of detailed information at the same fine resolution defined by the input data source is quite challenging. Therefore, large-scale data assembly capability was developed to generated aggregated simulation results across spatially-explicit landscape in a self-described format. Furthermore, interactive data visualization and exploration software utilities have been developed for those gridded data products over spatially-explicit landscapes. Based on those spatially-explicit data products, other environmental models, such as water quality models, biodiversity, as well as environmental impact model can linked together to provide a more comprehensive knowledge system for sustainable landscape management.

18.3 Landscape Is the Key Component Within Integrated Environmental Modeling

As presented in all of those three examples, one of the fundamental concepts for integrated environmental system modeling is to identify shared relationships between all subsystems of the modeled environment. Take the integrated model system for a urban environment as an example, the connections between some systems, such as the transportation system and regional air quality modeling, are explicit and direct. However, in the same environment, the relationship between the water quality treatment and transportation system not directly, but indirected connected. Considering the strong interests on the water-energy-transportation-social simulation, the spatially-explicit landscape became one of few direct links across all the model systems. There are different approaches to manage those data processing across spatially-explicit landscape, for example, a dedicated landscape library was developed in the ATLSS case, a GIS-based landscape was used in the sustainable urban infrastructure effort, and self-descried data format (netCDF and geoTIFF) was used in the agroecosystem modeling practices. From our experience a light-weighted, high performance georeferencing processing capability is one of the essential functions for the landscape component. Other key functions are regriding/projection between multiscale computational domains, basic IO functions, as well as point, polygon and raster overlay.

18.4 Integrated Environmental System Modeling Needs Modern Accelerator Technologies

This section reviews current environmental modeling literature which adopted modern accelerator technologies with focus on general purpose graphic processing units and multicore system.

18.4.1 Spatially-Explicit Environmental Applications Using Graphical Processing Unit

Agent parallelism approaches provide unique features for the large scale of spatial-explicit agent-based models (Parry and Evans 2008). One is the agent-parallel method that splits agents between cores. It is able to decompose functions or processes and executed on different cores. Some successful applications of the agent-parallel method include a series of ecological agent-based models, such as aphids and hoverflies (Parry and Evans 2008), Tress (one processor for each tree; Host et al. 2008), and landscape vegetation model (Cornvell et al. 2001). This type of application can directly balance load across cores, but need to pay attention on spatial co-location in space. The other is the environment-parallel approach that can divide geographical location between cores. The grid cells are assigned for different cores. The advantage of this method is that cores associated with agents are independent, and easy to manage or track. Forest modeling (Chave 1999), disaster mitigation (Takeuchi 2005) and Land Use Change (Tang et al. 2011) are the application examples of the environment-parallel design of agent-based models.

GPU application provides a promising tool for spatial explicit environmental modeling. A framework of personal high-performance geospatial computing (HPC-G) was proposed to use multi-processors CUPs and massively parallel GPU devices for large scale of geospatial data processing (Zhang 2010). This framework with limited costs is suitable to both numerical modeling and interactive visualization. Bryan (2013) explored how different high-performance computing tools with multi-nodes, multi-core and GPU. Compared with single-machine GIS application, the 64 parallel GPU application demonstrated a speed-up of 63,643 fold for 250-m-resolution national social-ecological modeling (~ 100 million simulation units) in Australia, demonstrating a high potential than other HPC designs.

Because of the enormous computing power, GPU also provides new opportunities for uncertainty quantification and sensitivity analysis of environmental applications. However, the new hardware architecture generally requires large effort to related to software re-engineering of existing environmental applications. From this perspective, it will take time for the environmental modeling community to take the GPU as a general purpose computing platform as the multicore systems, but authors are confident that GPU will be featured in many environmental modeling areas, such as those typical spatially-explicit environmental applications, mentioned in this paper.

18.4.2 Environmental Applications Using Multicore Technologies

Over the past decades, multicore technologies have been adopted to expedite the large-scale environmental applications, including water resource modeling (Yu 2010), air quality modeling (Matthias Lieber1 and Wolke 2008), climate modeling (Wang et al. 2011), and ocean modeling (Chen et al. 2006). More recently, integrated modeling was emphasized, resulting in several system modeling (Collins et al. 2005) and computational (both simulation and data processing) frameworks (Bader et al. 2011; Wilbert et al. 2011). There are also reports on the high performance modeling architecture design (Collins et al. 2005; Bader et al. 2011) and the coupling component implementations (Wang et al. 2011; Larson 2005). From the environmental modeling perspective, a meta-modeling framework was presented for large-scale, spatially-explicit ecosystem modeling on computing grids (Wang et al. 2005a). High performance geographic information systems have been used to handle large environmental and ecosystem information (Wang et al. 2006; Yin et al. 2011).

Embarrassingly parallel design with multi-nodes or multi-cores has also been applied to high-resolution ecosystem modeling through scientific workflow systems under clusters or distributed computing grids. Most early ecosystem or environmental system models were designed sequentially in last century, and are not able to directly fit current parallel systems for a large scale of modeling. Redesigning and coding these models for parallel application together with testing would be very time consuming. The embarrassingly parallel computing by separating tasks for individual spatial explicit modeling unit provides an efficient tool of these traditional sequential environmental models. Nichols et al. (2011) developed a scheme to execute 120,000 simulations for 56-m resolution of environmental modeling of agricultural ecosystems (EPIC) on a cluster within 8 h. In a high-resolution of landscape and water quality study of Europe, a sequential watershed model, soil and water assessment tool (SWAT) was split into major sub-models, and executed in parallel on the Enabling Grids for E-scienceE (EGEE) Grid (Yalew et al. 2010). The initial test indicated that the SWAT model can reach speed-up of about three with seven CPUs for the Balaton Lake watershed. A hybrid approach distributes tasks across a heterogeneous grid simulated 325 management scenarios (nitrogen application rates and stubble management) at a daily time step over 122 years for 12,707 units using the Windows-based Agricultural Production Systems SIMulator (APSIM) (Zhao et al. 2012). These simulations would have taken over 30 years on a single computer. The hybrid high performance computing (HPC) approach based on embarrassingly parallel design completed the modeling within 10.5 days-a speed-up of over 1,000 times. High performance visualization tools have also been developed to display a three-dimensional environment for farming system analysis (Gaff et al. 2000). A successful application of improved hydraulic modeling based on OpenMP designed by Neal et al. (2008) was tested on different number of cores of machines. The results show that this improvement facilitates high-resolution modeling (millions of cells) for a large basin such as Amazon in Brazil.

18.5 Opportunities of Adopting Accelerator Technologies for SIEM

In the previous sections, we have reviewed some literatures on adopting modern accelerator technologies for environmental system modeling. There are many reports on the application of GPU on both agent-based modeling and environmental model implementations. Herein, authors layout several computational opportunities in the general context of spatially-explicit environmental system modeling, which can directly take advantages of the modern accelerator technologies. Here three areas are listed, including (1) data management for multiscale environmental data processing over landscape, (2) social-environmental system interactions, and (3) global sustainable landscape management scenarios development.

18.5.1   High-Performance Data Management to Support Multiscale Data Processing over Landscape

A central issue in integrated environmental modeling is the need to link dynamic models that operate across different spatial regions and at different rates. Therefore, the data and information processing over landscape component is the key component of spatially-explicit, integrated environmental modeling. The main functionalities of the landscape component include (1) provide a uniform structure to integrate spatial information and information about physical data (such as flux and energy) from different model components; (2) generate appropriate computational domains (meshes) for different models; (3) conduction conservative mapping of energy and mass between the different computational domains; and (4) determine parameter estimates for environmental models at different spatial resolution.

One possible solution could be the development of a standalone landscape libraries to transform and translate the spatial data to and from other forms. Universal transect mercator (UTM) is widely adopted in GIS, and could be used to extract regional information from geo-spatial data. In a GIS, regions are usually based on the polygon's shape and location, defined in terms of UTM coordinates. It's therefore possible to extract the same regional information from data sets with different resolutions, registrations, or spatial extents. Based on the UTMs, re-projections and regrid function can be developed to facilitate data transformation between different resolutions and registrations. Users can configure and initialize these Regrid classes by providing the size of a single cell or assigning the exact number of rows and columns to which the data sets should be resized. Spatial data sets can then be passed to the object, which creates and returns a new, rescaled map. Other important functions, which are frequently used in environmental data analysis, include Zonal Statistics for summarizes a raster/point layer based on the zones defined in a polygon layer. One promising approach could be to develop a novel data organization scheme to virtually integrate heterogamous data sources and another is a set of parallel algorithms to support the important operations based on the data organization scheme using modern accelerator technologies. These algorithm could be viewed as the geospatial indexing and processing engines to speed up data processing without requiring significantly reorganizing raw data. Technically, we could implement the software modules using the techniques actively scan relevant data and generate different types of indices to regularize workloads for efficient parallel computing.

18.5.2 Large-Scale Social-Environmental System Interactions via Agent-Based Modeling

Large-scale agent based models could be used to help understand the effects of micro-level consumer psychological factors (e.g., attitudes, preferences, satisfaction, and behavior) and social networks where consumers interact on the macro level innovation diffusion processes via dynamic simulation methods. How the impact of government policies and regulations on the dynamics of innovation diffusion can be modeled at microscopic level (e.g., the behavior and interactions of firms and adopters). Take the sustainable urban infrastructure project as an example, sustainable homes and neighborhoods are about incorporating green technologies to design and operate homes, buildings and neighborhood infrastructures. Green technologies cover a broad set of creative inventions including renewable materials, distributed energy production, rainwater harvesting, water-saving facilities, low impact development, and so on. As a novel strategy to create more sustainable and resilient living environment, sustainable homes and neighborhoods are still in its infancy and the wide adoption among home developers and homeowners will be a long and slow process. Agent based model can be developed to predict and compare the innovation diffusion processes under different policies and strategies. This evaluation can help policy makers and stakeholders find the most effective policies and strategies in fostering the diffusion of sustainable homes and neighborhoods.

Agent based model could be developed to understand the social decision making that is involved in increasing the adoption of more sustainable homes and neighborhoods, and how we can improve citizen capacity building to create more informed citizens. This understanding represents the beginning step to understand the socio economic drivers that would be required for building more sustainable cities and infrastructure. After all, citizens make decisions two feet at a time in the shopping aisle and voting booth and this is the driving force for more sustainable decision making. A computational infrastructure based on modern accelerator technologies could facilitate the assessing procedure how decisions by households and firms relate to sustainability outcomes for larger regions under different policies such as fees and incentives. Computational models could be created to examine the benefit of green urban infrastructure on water, material and energy use and return on investment for the following approaches to increasing sustainability: (1) decentralized, combined heat and power using air-cooled gas fired turbines, (2) electrification of personal transportation, (3) decentralized storm water and water production using low impact development, (4) increased common green space and higher population density.

18.5.3 Global Sustainable Landscape Management Scenario Development

The concerns over the urgent global issues (e.g. food security, bioenergy production and climate change impact) associated with agroecosystems call for examining and redesigning current ecosystems. These need detailed information of agroecosystem responses to various management practices. Spatially-explicit agroecosystem modeling frameworks with parallel design provide a unique capability of upscaling field simulations to regions and globe without losing details. A series of research projects (e.g. Global Sustainable Bioenergy (GSB) and Knowledge Systems for Sustainability (KSS)) have been conducting or are under planning to provide the best data and knowledge of global ecosystem management with large-scale biophysical simulations on the basis of multi-core parallelism techniques. Specifically, The Global Sustainable Bioenergy (GSB) project seeks to contribute to a sustainable world by expanding, disseminating, and applying understanding of the possibility and necessity of producing bioenergy on a very large scale (Lynd et al. 2011). However, critics off current bioenergy technologies in combination with extrapolation of current trends and practices see as negatively impacting food security, having an inadequate resource base to meaningfully impact energy-related challenges, and contributing to environmental degradation. A joint research initiated by multiple international institutions targets the issues of land competition and sustainability by implementing global high-resolution food and bioenergy crop simulations (Kang et al. 2012). A platform has been designed and tested for a bioenergy crop, switchgrass. Over 10 bioenergy crops and field-scale resolution modeling have been planning, which require redesigning current platform for adapting millions of simulations on supercomputers. For example, with the HPC-EPIC platform, 2.2 billion of simulations for global agroecosystem at a 30-s resolution (~ 1 km) under 10 management scenarios would be implemented under a cluster or supercomputer. This can not only facilitate detailed management optimization at a local scale, but also easily aggregate regional, national and global distributions of production and environmental impacts for multi-level of decision making.

18.6 Conclusions and Future Readings

In this section, authors have discussed three research areas, which can benefit directly from the further adoption of modern accelerator technologies. There are other areas, such as large-scale data exploration and information synthesis, will also benefit from accelerator technologies, but authors leave out those applications, because of the focus on chapter was placed on the spatially explicit integrated environmental modeling.

Integrated environmental system modeling is a promising practice to better understand interactions between the human and major components of environmental systems. Spatially-explicit modeling approach is a natural choice for those modeling activities, because landscape is one of few shared components across all models. In this chapter, authors explained the importance and role of spatially-explicit landscape in those efforts, and listed three potential research areas, within the scope of integrated environmental system model, which could be further developed in the near future using modern accelerator technologies. Interested readers should look into the references for further detailed information.

Acknowledgements Authors thanks the support from the Office of Science of the U.S. Department of Energy (DOE). Oak Ridge National Laboratory is managed by UT-Battelle LLC for the Department of Energy under contract DE-AC05-00OR22725.

References

AERMIC. (2004). "AERMOD atmospheric dispersion modeling system." from http://www.epa.gov/scram001/7thconf/aermod/aermod_mfd.pdf.

ATHENA. (2007). "Impact Estimator for Buildings." from http://www.athenasmi.org/our-software-data/impact-estimator/.

Adams, D. M., Alig, R. J., Callaway, J.M., McCarl, B. A., Winnet, S.M. (1996). The forest and agricultural sector optimization model (FASOM): model structure and policy applications. Department of Agriculture, Forest Service, Pacific Northwest Research Station: Portland, OR. p. 60.

Bader, M., Mehl, M., Rde, U., Wellein, G. (2011). Simulation software for supercomputers, Journal of Computational Science, 2(2), 93–94.

Barrett, C., Birkbigler, K., Smith, L., Loose, V., Beckman, R., Davis, J., Roberts, D., Williams, M. (1995). An Operational Description of TRANSIMS. Technical Report. Los Alamos, NM, Los Alamos National Laboratory.

Byun, D., Schere K.L. (2006). "Review of the governing equations, computational algorithms, and other components of the Models-3 Community Multiscale Air Quality (CMAQ) modeling system." Appl. Mech. Rev 59(51–77).

Bryan, B.A. (2013). High-performance computing tools for the integrated assessment and modeling of social-ecological systems. Environmental Modeling & Software, 39, 295–303.

Chen, C., Beardsley, R. C., Cowles, G. (2006). "An unstructured grid, finite-volume coastal ocean model (FVCOM) system." Oceanography 19(Special Issue entitled "Advance in Computational Oceanography): 78–89.

Collins, N., Theurich, G., DeLuca, C., Suarez, M., Trayanov, A., Balaji, V., Li, P., Yang, W., Hill, C, Silva, D. (2005). Design and Implementation of Components in the Earth System Modeling Framework. International Journal of High Performance Computing Applications, 3, 341–350.

Comiskey, E. J., Gross, L. J., Fleming, D. M., Huston, M. A., Bass, O. L., Luh, H.K., Wu,Y. (1994). A spatially-explicit individual-based simulation model for Florida panther and white-tailed deer in the Everglades and Big Cypress landscapes. Florida Panther Conference, Ft. Myers Fla, U.S. Fish and Wildlife Service.

Chave, J. (1999). Study of structural, successional and spatial patterns in tropical rain forests using TROLL, a spatially explicit forest model. Ecological Modeling, 124, 233–254.

Cornvell, C.F., Wille, L.T., Wu, Y.G., Sklar, F.F. (2001). Parallelization of an ecological landscape model by functional decomposition. Ecological Modeling, 144, 13–20.

Duke-Sylvester, S., Gross, L.J. (2002). Integrating Spatial Data into an Agent-Based Modeling System: Ideas and Lessons from the Development of the Across Trophic Level System Simulation (ATLSS). Integrating Geographic Information Systems and Agent-Based Modeling Techniques for Stimulating Social and Ecological Processes. H. R. G. (ed), Oxford Univ. Press: 125–136.

Delden, H. V., Gutirrez, E.R., Vliet, J. (2008). Xplorah, a multi-scale integrated land use model. International Congress on Environmental Modeling and Software, Barcelona.

EPANET. (2008). "Software That Models the Hydraulic and Water Quality Behavior of Water Distribution Piping Systems." from http://www.epa.gov/nrmrl/wswrd/dw/epanet.html.

EQUEST. (2010). "The QUick Energy Simulation Tool." from http://doe2.com/equest/index.html.

Edmonds, J., Wise, M., Pitcher, H., Richels, R., Wigley, T., MacCracken, C. (1997). An Integrated Assessment of Climate Change and the Accelerated Introduction of Advanced Energy Technologies. Mitigation and Adaptation Strategies for Global Change, 1, p. 311–319.

Fernando, H. J., Lee, S. M., Anderson, J., Princevac, M., Pardyjak, E., Grossman-Clarke, S. (2001). "Urban fluid mechanics: air circulation and contaminant dispersion in cities." Environmental Fluid Mechanics 1(1), 107–164.

Gross, L., D. DeAngelis (2002). Multimodeling: New Approaches for Linking Ecological Models. Predicting Species Occurrences: Issues of Accuarcy and Scale. P. J. H. J.M. Scott, and M.L. Morrison, Island Press: 471–476.

Gaff, H., DeAngelis, D. L., Gross, L. J., Salinas, R., Shorrosh,M. (2000). "A dynamic landscape model for fish in the Everglades and its application to restoration." Ecological Modelling 127: 33–52.

Hunt, J. D., Abraham, J.E., Ed. (2005). Design and implementation of PECAS: A generalized system for the allocation of economic production, exchange, and consumption quantities. Integrated Land-Use and Transportation Models: Behavioural Foundations. Oxford, UK:, Elsevier.

Host, G.E., Stech, H.W., Lenz, K.E., Roskoski, K., Mather, R. (2008). Forest patch modeling: using high performance computing to simulation aboveground interactions among individual trees. Function Plant Biology, 35, 976–987.

Klosterman, R. (1999). "The What if? collaborative planning support system." Environment and Planning B 26(393–408).

Kang, S., Nair, S.S., Kline, K.L., Nichols, J.A., Wang, D., Post, W.M., Brandt, C.C., Wullschleger, S.D., Singh, N., Wei., Y. (2012). Global simulation of bioenergy crop productivity: analytical framework and case study for Switchgrass. GCB Bioenergy (accepted).

Larson, J., Jacob, R., Ong, E. (2005). The Model Coupling Toolkit, A New Fortran90 Toolkit for Building Multiphysics Parallel Coupled Models. International Journal of High Performance Computing Applications, 19(3), 277–292.

Li, K., Zhang, P., Crittenden, J.C., Guhathakurta, S., Chen, Y., Fernondo, H., Sawhney, A., McCartney, P., Grimm, N., Kahhat, R., Joshi, H., Knjevod, G, Choi, Y., Fonseca, E., Allenby, B., Gerrity, D., Torrens, P. (2007). "Development of a Framework for Quantifying the Environmental Impacts of Urban Development and Construction Practices." Environmental Sciences and Technologies 41(14), 5130–5136.

Lobell, D.B., Schlenker, W., Costa-Roberts, J. (2011). Climate trends and global crop production since 1980. Science, 29, 616–620.

Lynd, L., Aziz, R., Cruz, CH, et al. (2011) A global conversation about energy from biomass: The continental conventions of the global sustainable bioenergy project. Interface Focus, 1, 271–279.

Makropoulos, C. K., Natsis, K., Liu, S., Mittas, K., Butler, D. (2008). "Decision support for sustainable option selection in integrated urban water management." Environmental Modelling & Software 23(12): 1448–1460.

Matthias Lieberl, M., Wolke, R., Optimizing the coupling in parallel air quality model systems, Environmental Modeling and Software, Volume 23, Issue 2, 2008, pages 235–243

Nelson, A. (2004). Toward a New Metropolis: The Opportunity to Rebuild America. Washington, DC, The Brookings Institute.

Neal, J., Fewtrell, T., Trigg, M. (2008). Parallelization of storage cell flood models using OpenMP. Environmental Modeling & Software, 24, 872–877.

Nichols, J., Kang, S., Post, W., Wang, D., Bandaru, V., Manowitz, D., Zhang, X., Izaurralde, R. (2011). HPC-EPIC for high resolution simulations of environmental and sustainability assessment. Computers and Electronics in Agriculture, 79, 112–115.

Parry, H.R., Evans, A.J. (2008). A comparative analysis of parallel processing and super-individual methods for improving the computational performance of a large individual-based model. Ecological Modeling, 214, 141–152.

Trusty, W. B. (2004). Life cycle assessment, databases and sustainable building. Latin-American Conference on Sustainable Building. Sao Paolo.

Takeuchi, I. (2005). A massively multi-agent simulation system for disaster mitigation. In Massively Multi-agent Systems I: First International Workshop MMAS 2004. Kyoto Dec 2004. Heidelberg: Springer-Verlag.

Tang, W., Bennett, D., Wang, S. (2011). A parallel agent-based model of land use opinions. Journal of Land Us Science, 6, 121–135.

Thomson, A.M., Rosenbergy, N.J., Izauralde, R.C., Brown, R.A. (2005). Climate change impacts for the conterminous USA: an integrated assessment. Climate Change, 69, p. 27–41.

Voinov, A., Hood, R. R., Daues, J. D., Assaf, H., Stewart, R. (2008). Building a Community Modeling and Information Sharing Culture. Environmental Modelling, Software and Decision Support, Volume 3: State of the art and new perspective (Developments in Integrated Environmental Assessment). A. J. Jakeman, Voinov A. A., Rizzoli, A. E., Chen, S. H. Amsterdam, The Netherlands, Elsevier: 345–365.

VEMAP_Members, Vegetation/Ecosystem Modeling and Analysis Project: Comparing biogeography and biogeochemistry models in a continental-scale study of terrestrial ecosystem responses to climate change and CO_2 doubling. Global Biogeochem. Cycles, 1995. 9(4): p. 407–437.

Waddell, P. (2000). "A behavioral simulation model for metropolitan policy analysis and planning: residential location and housing market components of UrbanSim." Environmental Planning 27(2), 247–263.

Wang, D., Berry, M.,W., Gross, L.J. (2008). "A Parallel Structured Ecological Model for High End Shared Memory Computers." Proceedings of the First International Workshop on OpenMP 4315.

Wang, D., Berry, M.W., Carr, E. A., Gross, L.J. (2006). "On Parallelization of a Spatially-Explicit Structured Ecological Model." International Journal on High Performance Computer Applications: 571–581.

Wang, D., Berry, M.W., Buchanan, N., Gross, L.J. (2006). A GIS-enabled Distributed Simulation Framework for High Performance Ecosystem Modeling. ESRI International User Conference.

Wang, D., Berry, M.W., Comiskey, J., Gross, L.J. (2007). A Parallel Simulation Framework for Integrated Regional Ecosystem Modeling. The 2007 International Conference on Parallel and Distributed, Processing Techniques and Applications PDPTA'07.

Wang, D., Carr, E., Gross, L.J., Berry, M. W. (2005). "Toward Ecosystem Modeling on Computing Grids." IEEE Computing in Science and Engineering(Sep/Oct): p 44–52.

Wang, D., Carr, E.A., Berry, M.W., Gross, L.J. (2005). "A Grid Service for Natural Resource Managers." IEEE Internet Computing(Jan/Feb): pp 35–41.

Wang, D., Carr, E.A., Berry, M.W., Gross, L.J. (2006). "A Parallel Fish Model for Ecosystem Modeling." Simulation: Transactions of The Society of Simulation and Modeling International(July): 451–465.

Wang, D., Harmon, M., Berry, M.W., Gross, L.J. (2011). On Design of a Coupling Component for Parallel Multimodeling. International Journal on Modeling, Simulation, and Scientific Computing.

Wang, D., Post, W., Wilson, B. (2011). Climate Change Modeling: Computational Opportunities and Challenges, IEEE Computing in Science and Engineering, 13(5), 36–42.

Wolff, W. F. (1994). "An individual-oriented model of a wading bird nesting colony." Ecological Modelling 72, 75–114.

Wang, D., Kang, S., Post. W., Nichols, J., Zhao, Z., Liu, S. (in press) A Computational Framework for Spatially-explicit Agroecosystem Modeling: Application to Regional Simulation, Journal of Computational Sciences, DOI:10.1016/j.jocs.2012.08.018

Wilbert, N., Zito, T., Schuppner, R., Jedizejewski-Szmek, Z., Wiskott, L., Perkes, P. (2011). Building extensible frameworks for data processing: The case of MDP, Modular toolkit for Data Processing, Journal of Computational Science, (in press) Available online 29 October 2011, http://dx.doi.org/10.1016/j.jocs.2011.10.005

Yalew, S.G., Griensven, A.V. (2010) Kokoszkiewicz, L. Parallel computing of a large scale spatially distributed model using the soil and water assessment tool (SWAT). International Environmental Modeling and Software Society(iEMs), International Congress on

Environmental Modeling and Software Modeling for Environment's Sake, fifth biennial meeting.

Yin, L., Shaw, S.L., Wang, D., Carr, E., Berry, M., Gross, L., Comiskey, J. (2011). A problem solving framework of integrating GIS and parallel computing for spatial control problems: A case study of wildfire control. International Journal of Geographic Information Science, Vol 26, December. p 1–21. DOI: 10.1080/13658816.2011.609487

Yu, D. (2010). Parallelization of a two-dimensional flood inundation model based on domain decomposition, Environmental Modelling and Software, 25(8), 935–945.

Zhao, G., Bryan, B.A., King, D., Liu, Z., Wang, E., Bende-Minchl, B., Song, X., Yu, Q. (2012). Large-scale, high-resolution agricultural systems modeling using a hybrid approach combining grid computing and parallel processing. Environmental Modeling & Software 10, http://dx.doi.org/10.1016/j.envsoft.2012.08.007

Zhang, J. (2010). Towards personal high-performance geospatial computing (HPC-G): perspectives and a case study. HPDGIS' 10 Proceeding of the ACM SIGSPATIAL International Workshop on High Performance and Distributed Geographic Information Systems. doi:10.1145/1869692.1869694

Printed in the United States
By Bookmasters

Printed in the United States
By Bookmasters